"Professor Quan Zhang is one of the major scholars in language testing and Rasch measurement theory in China. This book provides a clear introduction to both classical test theory and modern measurement research on the Rasch model. Quan is a Board member of the Pacific Rim Objective Measurement Society. He is the author of Rasch-GZ software that provides the opportunity for students and others to apply the measurement principles presented in this book. Rasch-GZ is the first Chinese platform for both CTT-based Item Analysis and Rasch-based Test Equating. This book will be of help to scholars with an interest in using and improving measurement practices in the human sciences".

George Engelhard, *Recipient of Benjamin Drake Wright Senior Scholar Award (2018), Georgia University, USA*

"Quan Zhang has had a long-term interest in language testing. He has played a central role in developing the Pacific Rim Objective Measurement Symposia, particularly in mainland China. This puts him in an ideal position to write the book".

Trevor Bond, *Recipient of Benjamin Drake Wright Senior Scholar Award (2025), Lead author "Applying the Rasch Model"*

"This six-chapter monograph on Rasch modelling for language testing is a must not only for test developers but also for all serious researchers in the field of assessment and evaluation. The book presented a well-structured introduction to the history and development of Rasch models, and a scholastic comparison between the classical model and the Rasch approach. It then gave detailed discussions on how to set up the analysis and interpret the outputs. I appreciate especially the author's use of concrete and interesting test development cases to highlight critical points in the process. The book also provided important resources including the PROMS conferences and workshops in support of the readers' further development. I am very much impressed by the breadth of scope, and the depth in intensity and thoroughness. I can recommend this book to postgraduate students and researchers who aspire to rigor and validity in their research findings".

Magdalena Mo Ching Mok, *The Education University of Hong Kong, Hong Kong, SAR, China*

"This book is an insightful contribution to the field of language assessment. It bridges the gap between complex statistical model and practical testing scenarios. It serves as an invaluable resource for researchers and practitioners aiming to enhance the quality of their assessments through item response theory".

Xiaoting Huang, *Peking University, China*

"I believe this book will help readers to bridge the gap between theoretical frameworks and practical applications. This book also introduces Rasch-GZ, a cutting-edge software that integrates AI-driven assistance and bilingual support for language testing. With Rasch-GZ, readers can access supplementary video tutorials in English, facilitating hands-on learning and software mastery to address challenges in language learning and testing".
A.Y.M. ATIQUIL ISLAM, *the East China Normal University, Shanghai, China*

"*Rasch Model for Language Testing* integrates theoretical perspectives with practical applications. This book will be a valuable reference for both researchers and practitioners seeking to address real-world measurement challenges in language testing".
Zi Yan, *The Education University of Hong Kong, Hong Kong, SAR, China*

"Professor Zhang's book is a valuable addition to the Rasch measurement literature, making a substantial contribution to the field. It provides readers with a comprehensive introduction to the Rasch model, extending to advanced topics such as test equating. Notably, the book also includes a dedicated chapter on PROMS conferences focused on advancing Rasch measurement".
Bambang Sumintono, *Universitas Islam International, Indonesia*

"Combining theoretical rigor with real-world application, this book deepens our understanding of Rasch measurement and its relevance across educational and social science contexts".
Che Yee Lye, *SUSS, Singapore*

"This book introduces both CTT and the Rasch model in an integrated approach, providing a comprehensive pathway to enhanced measurement accuracy and efficiency in the context of language testing and bridging the gap between theoretical frameworks and practical applications".
LUO, Guanzhong, *South China Normal University, China*

Rasch Model for Language Testing

This book focuses on the crucial theme of modern test development and analysis, specifically focusing on bridging the gap between theoretical frameworks and practical applications. This resource offers a practical and detailed exploration of psychometric principles, encompassing both classical test theory (CTT)-based item analysis and item response theory (IRT), specifically focusing on the one-parameter model (Rasch model) for test equating. By considering CTT and the more sophisticated Rasch model, the book provides a comprehensive pathway to enhanced measurement accuracy and efficiency in the human sciences. It aims to empower readers with the tools and knowledge necessary to navigate the complexities of test equating, item bank creation, and the implementation of computerized adaptive testing (CAT). It provides historical context for these methodologies and outlines prospects for Rasch research, emphasizing the importance of establishing a common scale for measuring ability.

A significant contribution is the introduction of Rasch-GZ, a cutting-edge software developed during the COVID-19 pandemic that incorporates AI-driven assistance and bilingual support, potentially streamlining the calibration process of items within an item bank. The book thoroughly reviews fundamental concepts such as item difficulty and discrimination, ensuring readers grasp the theoretical underpinnings of both CTT and IRT. It delves into various test equating techniques, including Common Item Linking and potentially Common Subject Linking, crucial for maintaining invariance across different test forms within a Standardized Examination. Real-world examples with detailed interpretations serve as a step-by-step guide to these procedures, often relying on anchor items. While implicitly addressing the concept of the norm in the context of standardized examination interpretation, the resource primarily focuses on the application of psychometric software. Access to the student version of Rasch-GZ is provided through code, alongside supplementary video tutorials, facilitating hands-on learning and software mastery.

This book could be used by practitioners in the fields of applied linguistics, applied psychology, language testing, education management, and researchers working on language testing methods.

Quan Zhang was Dean of English College of several universities in China, has research interests from linguistics to statistics with expertise on language testing, Rasch-based test equating, and quantitative measurement. He used to be the key member of MET equating project sponsored by the MoE, PR China. As a representative of Pacific Rim Objective Measurement Symposium (PROMS), he has organized several international conferences and has many publications.

Rasch Model for Language Testing
Advances of PROMS and RASCH_GZ with Use Cases

Quan Zhang

CRC Press
Taylor & Francis Group
Boca Raton London New York

CRC Press is an imprint of the
Taylor & Francis Group, an **informa** business
A CHAPMAN & HALL BOOK

Designed cover image: Getty

First edition published 2026
by CRC Press
2385 NW Executive Center Drive, Suite 320, Boca Raton FL 33431

and by CRC Press
4 Park Square, Milton Park, Abingdon, Oxon, OX14 4RN

CRC Press is an imprint of Taylor & Francis Group, LLC

ISBN: 978-1-041-12596-9 (hbk)
ISBN: 978-1-041-12595-2 (pbk)
ISBN: 978-1-003-66559-5 (ebk)

DOI: 10.1201/9781003665595

Typeset in Palatino
by KnowledgeWorks Global Ltd.

Contents

Preface

This book is written to promote the research and application of the Rasch model for item analysis and test equating. It is divided into six chapters: Chapter 1 traces the research background, the basic concepts with the terminology, the origin of the Rasch model, Rasch people, and the events related to the Rasch model study, addressing the differences and relationships between the Rasch model and CTT, as well as between the Rasch model and the one parameter model of IRT; Chapter 2 provides a brief review of the ABC knowledge of CTT used for item analysis in language testing. These are important concepts for new comers. Chapter 3 discusses the specific application of test equating based on the Rasch model, the concepts and significance of test equating, both the use cases and conditions for conducting test equating with a simplified example, and introduces one by one the recognized Rasch-based and IRT-based software packages for test equating in the language testing field with the comments based on the author's personal experiences. Chapters 4 and 5, respectively, provide the interface and specific operating steps for CTT-based item analyses and the Rasch-based test equating of the latest Rasch GZ system developed by the author in the AI era. Chapter 6 gives information about Pacific-Rim Objective Measurement Symposium (PROMS), introducing the development of PROMS conferences since the first conference held in Malaysia in 2005 with detailed description regarding the pre-conference workshops and keynote speeches of each conference over the past 20 years. In particular, the author reminds the readers that the rapid growth of research and application in psychology, education, management, linguistics, medicine, sports and other disciplines in the humanities and social sciences has further shown us the broader prospects of Rasch measurement research and promotion, as well as the value of the author's persistent research and efforts made to promote the study.

Upon the publication of the book, thanks are given to the strong support of Taylor & Francis CRC Publishing House to have the book published; thanks also go to City Training Institute (CTI), Guangzhou and Guangzhou Art & Science Technology Development Co., Ltd. for their supports.

Meanwhile, I have a lot of people to thank, among whom I do want to mention the names of:

Prof. A.Y.M. Atiquil Islam who gave me lots of advices regarding book writing; Prof. Trevor Bond who wrote in person the information about his position in the Rasch family, and the paragraph about the first PROMS 2005, Kuala Lumpur, Malaysia; Prof. Luo Guanzhong, as a RUMM program developer, who gave me the exact information about RUMM; Prof. Magdalena Mo Ching Mok who provided me the information about VIPs in the Rasch family;

Prof. Bambang Sumintono who kindly provided me with all the information about PROMS2019, Surabaya, Indonesia; Prof. Mohd Nor Mohd Zali who supplied all the information about use cases of the Rasch Model in Malaysia. Finally, lovely appreciation goes to Prof. Xiaoxi Hu, my wife who proofread the book from cover to cover and gave me a lot of suggestions.

While they saved me from a number of inaccuracies and infelicities, they can in no way be held responsible for the academic opinions which are expressed and any imperfection and shortcomings which may remain.

The author
From Conghua Hot Spring Hotel
Guangzhou, China, June, 2022

Glossary

A

AALA: Asian Association for Language Assessment

α value: The Cronbach's alpha coefficient value

Abnormal distribution: (Cf. normal distribution)

ACARA: Australia Curriculum, Assessment and Reporting Authority

Achievement Test (AT): A textbook-based test, used to measure what students have actually learnt from a textbook based on teaching syllabus over a period of time.

ACER: Australian Council for Educational Research

ACT: American College Test in the United States

Additivity: Ability or motivation on a linear scale. See invariance, linear measurement, log-odds units, separation of person and item parameters, and summation property.

AES: Automated Essay Scoring

A-Level: Advanced Level, a qualification offered in the UK and part of the General Certificate of Education considered as a prerequisite for university entry

AERA: American Education Research Association

AMOS: Analysis of Moment Structures

Anchored items: Different term for common items (cf. Common item linking)

ANOVA: Analysis of variance

APHA: American Public Health Association

APT: Association of Test Publishers

ASEAN: Association of Southeast Asian Nations, an international organization established in 1967 by Indonesia, Malaysia, the Philippines, Singapore, and Thailand to promote economic growth, peace, and security in Southeast Asia. Brunei joined in 1984, followed by Vietnam (1995), Laos and Myanmar (1997), and Cambodia (1999). East Timor has observer status with expectations to become a full member in 2025

B

Bartlett: Bartlett's test of sphericity, the test introduced by the English statistician Maurice Stevenson Bartlett in 1937 and used to assess whether a correlation matrix is significantly different from an identity matrix (cf. KMO)

BEAR: Berkeley Evaluation and Assessment Research

BIGSTEPS: One of the earlier software programs developed for conducting Rasch analysis. The software handles both dichotomous and polytomous data, making it versatile for various types of assessments. While BIGSTEPS was widely used in the past, it has largely been superseded by Winsteps (cf. Winsteps)

BILOG: Item Analysis and Test Scoring with Binary Logistic Models, one of the representative software based on the IRT and originally developed by Prof Robert J. Mislevy of the University of Maryland and R. Darrell of the University of Chicago, USA

Biserial correlation coefficient: A statistical measure used to evaluate the quality of test items, particularly in the context of item analysis of language testing; such a correlation is used to determine how well a test item discriminates test-takers from high ability to low-ability

C

CAHSGE: California High School Graduation Examination in the United States

Calibration: The process of converting raw scores to logits on a Rasch measurement scale; item difficulty estimation, person ability estimation, fit analysis, scale development and iterative process

CAT: Computerized Adaptive Testing

CET: College English Test, designed for students of non-English major and implemented by the Education Examination Authority under the Ministry of Education, China. CET scoring report is based on national norm

CFA: Confirmative Factor Analysis (Cf. Mplus)

Chi Square test: Also X^2 Test, a statistical hypothesis-testing method used to compare observed frequencies with expected frequencies, helping determine whether there exists a significant difference between the expected and observed data under the None hypothesis. The key aspects of the Chi-square test include: Null and alternative hypotheses, P-value, goodness of fit test, and test of Independence

Classical Test Theory (CTT): One of the traditional theories dating back to the early 20th century in the assessment of psychological and educational tests, providing essential tools for understanding and improving the reliability and validity of tests; a dominant framework in psychometrics, guiding the formation of standardized tests in educational settings (cf. True score theory)

Common item linking (CIL): A kind of equating practice that the same set of well calibrated test items put in two separate test papers

Common subject linking (CSL): A kind of equating practice that two different test papers are administered to the same group of sampled test takers

ConQuest: One of the representative Rasch-based software developed by Mark Wilson

Convergence: When iteration reaches a pre-determined value, the iteration cycles stop, i.e., it converges (cf. Iteration)

Correlation: A certain degree of connection between variables; the covariation between variables, where one variable changes with the variation of another variable; the strength and direction of the relationship between the scores of the same test takers in two tests

COVID-19: Corona Virus Disease

Criterion-referenced test (CRT): A target-referenced exam. The practice of a test abides by a predetermined scale or goal when grading; if the scores given in NRT are relative, the scores given in CRT are fixed, meaning CRT does not take into account what scores achieved by other test takers in the same test taker population. See norm-referenced test

CTI: City Training Institute

D

Δ value: Represented by the Greek letters Δ (uppercase) or δ (lowercase) to indicate the difference or change between two variables. Originally a scale recommended by Educational Testing Service (ETS) in the United States: $\Delta = 13 + 4Z$

DELTA: Diagnostic English Language Tracking Assessment, a computer-based online assessment system based on Rasch measurement principles developed by the English Language Centre at Hong Kong Polytechnic University

Dichotomous scoring: Binary, two-valued ratings (right/wrong, yes/no, present/absent)

Discrimination: One of the index of item analyses which describes the degree to which an item is consistent with the other items in helping distinguish the proficiencies of the persons

Distractor: Option (not key answer) used in multiple-choice question type to produce some distracting effect

DSE: Diploma of Secondary Education, an entrance examination developed in Hong Kong, China

E

ECAE: Emirates College for Advanced Education in Abu Dhabi, UAE

EdSphere: An educational technology platform (also a brand name) designed and developed by Next Education India Pvt. Ltd. to enhance learning experiences by providing personalized and adaptive learning solutions. The name EdSphere likely combines "Ed", short for education, and "Sphere", suggesting a holistic approach to educational solutions

EdUHK: The Education University of Hong Kong

EFA: Exploratory Factor Analysis (Cf. Mplus)

EIRM: Explanatory item response models

EOCEP: End of Course Examination Program in the United States

EPT: English Proficiency Test, the predecessor of WSK, designed and implemented by the Education Examination Authority (formerly the

Examination Center of the Ministry of Education) under the Ministry of Education, China

Equating: A statistical process used to ensure that scores on parallel versions of a test are comparable, which is to seek for fair comparisons among different test-takers regardless of which version of the test they take and of what time to take

ETS: Educational Testing Service, Princeton, USA

External common item: A set of test items used to link parallel test forms but administered separately from the main test forms in the practice of test equating. These items are actually part of a separate test given to a sample of test-takers, providing a basis for equating scores across different test forms, administered to different groups of test-takers over the time (cf. Internal common items)

F

Facility: Level of difficulty; also facility index or facility value indicating the proportion of test-takers who answered a test item correctly

G

Gaokao: National College Entrance Examination in China

GEPT: General English Proficiency Test, a criterion-referenced testing system tailored for English learners in Taiwan, China

GITEST: The earliest Rasch-based software developed by Gui Shichun and Li Wei in China for CTT-based item analysis and Rasch-based test equating

Goodness of fit: The degree of matching between the actually observed data and the predicted results

H

HE: High education

Heterogeneity: The presence of variability within a dataset or among a population in descriptive statistics, indicating that the groups being compared exhibit diversity in some specific characteristic

High-stakes examinations: Examinations whose results are seen—rightly or wrongly—by students, teachers, administrators, parents, or the general public, as being used to make important decisions that immediately and directly affect them (Madaus as cited in Dianne Wall, 2005)

HKALE: The Hong Kong Advanced Level Examination

HKCEE: The Hong Kong Certificate of Education Examination

HKDSE: The Hong Kong Diploma of Secondary Education Examination

HKEAA: Hong Kong Examinations and Assessment Authority

HKiEd: Hong Kong Institute of Education, Hong Kong, SAR, China

Homogeneity: The similarity or uniformity within a dataset or among a population in descriptive statistics, indicating that the groups being compared are similar in some specific characteristic

I

IA: Item analysis

IACAT: International Association of Computerized Adaptive Testing (The University of Twente, the Netherlands)

ICEMEA: International Conference on Educational Measurement, Evaluation and Assessment

IELTS: International English Language Test System

IIUM: International Islamic University of Malaysia

Inter-rater inconsistency: The variability or lack of agreement in assessments or scoring made by different raters with different subjective judgment involved (cf. Intra-rater inconsistency)

Intra-rater inconsistency: The variability or lack of agreement in assessments or scoring made by the same rater at different times due to different factors

Invariance: The parameters of item difficulty are independent of the sample, and the estimates of subject ability are independent of the sample of items under certain conditions wherein the model fits the data well

IOMW: International Objective Measurement Workshop

Item and Person Parameters: Item difficulty and subject ability estimate

Item analysis: The core method in CTT refers to the internal evaluation of test items

Item pool: Also item bank, where a collection of calibrated test items is stored and where more equated items will be added. These items were administered to a small yet representative sampled group to evaluate their effectiveness, clarity, and appropriateness. The item pool is essential for ensuring the reliability and validity of CAT.

Item Response theory (IRT): A framework based on psychometrics to model the relationship between latent traits and performance on assessment items. The framework has three models: 1PL, 2PL, and 3PL

Item separation index (ISI): An index designed to indicate how the test items are spreading out across a range of difficulties, allowing for a more detailed analysis of the trait being measured. ISI (typically < 2.0) suggests a poor ISI, indicating that the test may not be reliably displaying between different levels of test item difficulties which could be centralized within a narrow range on a linear scale. See Person separation index

Iteration: The process of starting from an initial estimate, gradually approaching a fixed point. (Cf. Convergence)

Internal common item: A set of test items that are included in parallel test forms to serve as a basis for comparison in the practice of test equating. These items are used to link multiple versions of a test, allowing for the scores to be equated or compared across different administrations over time to different test takers (Cf. anchored items)

J

JAM: *Journal of Applied Measurement*
JLTA: Japan Language Testing Association
JML: Joint Maximum Likelihood

K

KMO: The Kaiser-Meyer-Olkin measure, a statistic used to assess the suitability of data for factor analysis, e.g., for a questionnaire design, evaluating the adequacy of the sample size and the degree of correlation among variables, with the range from 0 to 1, the closer to 1, the better the factor analysis. Typically,

KMO ≥ 0.90: Very good
0.80 ≤ KMO < 0.90: Good
0.70 ≤ KMO < 0.80: Above Average
0.60 ≤ KMO < 0.70: Average
0.50 ≤ KMO < 0.60: Poor
KMO < 0.50: Unacceptable

A KMO value of 0.60 or above is generally considered acceptable for proceeding with factor analysis (cf. Bartlett)

Kurtosis: Peak value, reflecting the concentration or flatness of the data distribution. (cf. Skewness)

L

Latent Traits: Attributes or (unobservable) properties that cannot be measured directly
Linear measurement: Raw scores being converted into measures on a linear scale on which the difference between any two points represents a consistent difference corresponding to the latent trait being measured, regardless of where (on the scale) the points are located. See additivity
LISREL: "Linear Structural Relations", a statistical software package used for SEM, path analysis, and other related multivariate statistical analyses
LLM: Large Language Modeling
Loading: The correlation coefficients between observed variables and latent factors, particularly in factor analysis. A loading indicates to what extend the variance in an observed variable can be explained by a factor. Factor loadings are crucial in interpreting the results, e.g., high loadings suggest that a factor strongly influences a particular variable, while low loadings indicate a weaker impact
Local independence: A key assumption in the Rasch measurement model and IRT models, underlying the idea that given a test-taker's ability level, the responses to individual test items are independent of each other
Local starting point: Local origin, the starting point of measurement; medium difficulty/ability on the measurement scale; initial estimates used for fitting model to data

Log$_e$: Abbreviated from "logarithm". Mathematically, natural logarithm (ln), with base $e = 2.718281828\ldots$ That is, $\ln(e^x) = x$, where e^x is the exponential function

Logit: Logarithmic odds unit transformed from raw scores via the Rasch model. The value of 0.0 logit is plotted at the center of the measurement scale, showing the M of an item difficulty or the subject ability estimate

Also a kind of representative software based on item response theory

Log-Odds: A logarithmic transformation of the odds to probability of success

Logistic Ogiv: The curve regarding the relationship between ability (relative to the test item) and the probability of answering it correctly. Logarithmic difference $(B_n - D_i)$

LSA: Latent Semantic Analysis

M

MCQ: Multiple choice question

MDI: Measurement-Driven Instruction introduced by Popham (1987)

Mean: The average usually represented by (M). In CTT, M is calculated by adding up the scores of all test takers and dividing by the number of test takers

MET: Matriculation English Test, one of the entrance examinations for higher education designed and implemented by the Education Examination Authority (formerly the Examination Center of the Ministry of Education) under the Ministry of Education, China

MFRM: Many Facet Rasch Model

ML: Maximum Likelihood. One of the most commonly used methods in estimating the parameters of a statistical model by maximizing the likelihood function, which measures how well the model explains the observed data

Model-data fit: The outcome regarding how well a statistical model represents the data it is intended to describe

MPA: Malaysian Psychometric Association

Mplus: A statistical software package developed by Linda Muthén & Bengt Muthén and used for SEM, EFA and CFA and other advanced statistical analyses

MUET: Malaysian University English Test.

N

NEEA: National Education Examinations Authority, China

Norm: Reference to reflect the scores of test takers of the homogeneous population

Normal distribution: A type of continuous, random variable, probability distribution; a normal (curve) distribution; a type of frequency distribution with three characteristics: symmetrical bell shaped curve characterized by the mean as the axis of symmetry, with peak values corresponding to the mean. A unimodal distribution

Norm-referenced test (NRT): A test taker's score with the scores achieved by other test takers for reference, statistically expressed in mean (M) and standard deviation (SD)

NSF: National Science Foundation, USA

O

One-Parameter Logistic Model (1PL): This model uses one parameter, i.e., the difficulty parameter (b) only, assuming all test items have the same discrimination and no guessing

P

Parallel tests: In language testing, two or more tests designed to be equivalent in form and in number of test items, the goal of which is to ensure that they measure the same constructs or they are to be equated and the equated tests can be used interchangeably

PARSCALE: IRT Item Analysis and Test Scoring for Rating Scale Data originally developed jointly by Dr Eiji Muraki at ETS and Professor R. Darrell Bock of the University of Chicago

Point Biserial correlation coefficient: A statistical measure used to assess the relationship between the overall test score and a dichotomous variable. A higher point-biserial correlation indicates that a particular item is well designed capable of discriminating test takers who perform well on the test are more likely to answer the item correctly, a low correlation suggests the item may not be functioning well; contrast to both, a negative correlation implies test takers who perform well on the test are more likely to answer the item incorrectly or those who perform poorly on the test are more likely to answer the item correctly (cf. Biserial correlation coefficient)

Person separation index (PSI): An index designed to discriminate between high and low ability of test takers. PSI (typically > 2.0) suggests a good ISI. See Item separation index

Pre-test: The practice of administering a test paper to a sampled target population to collect data for item analyses

Probability: Frequency of occurrence. The proportion of times the event would occur if repeating many times. The probability is always between 0 (never happens) and 1 (always happens)

Proficiency Test (PT): A test used to measure whether a test taker has reached level X, in order to determine whether they are suitable for a future task Y. The proficiency test takes into account what the students have learned (or should have learned)

PROMS: Pacific-Rim Objective Measurement Symposium; also Pacific-Rim Objective Measurement Society

P-value: The correct answer rate in CTT or the ratio of test takers who answered a question correctly to all the test takers in a test (R/T)

Q

QASS: Quantitative Applications in the Social Sciences

Qualitatively ordered data: Data ranked based on qualitative attributes but the intervals between the ranks are not necessarily quantifiable. A Likert scale is a typical example where responses such as "strongly agree", "agree", "neutral", "disagree", and "strongly disagree" are ordered, but the difference between each level is not precisely measurable. In Rasch measurement, the idea refers to data organized according to a qualitative hierarchy, typically in the form of ordinal categories used to measure latent traits or abilities

QUEST: The Quebec User Evaluation of Satisfaction with Assistive Technology

R

Range: The difference between two extremes; the full range can only be used as a rough indicator to infer the data distribution

Rasch-GZ: CTT-based Item Analysis and Rasch-based Test Equating System and online platform developed (2019–2023) by Quan Zhang, the author of the book. The first Rasch-based software launched in China that provides multiple-lingual interface and generate outputs in either English or Chinese subject to users' research background

Rasch model: The first psychometric model proposed by Danish mathematician Georg Rasch (1901–1981) for analyzing data from assessments, particularly for human sciences

Raw data: Unprocessed and un-edited data mostly containing errors, duplicates, or inconsistencies before being calibrated or analyzed

Raw scores: The scores obtained directly by a test taker on a test, showing the number of test items or questions answered correctly or the total points gained according to the rubrics

Reliability: Consistency. See True Score Theory

Residual: The discrepancy between an observed value (y) and the value predicted by the model $(\hat{y})$, i.e., $R = y - \hat{y}$

RMT: Rasch model theory, also Rasch measurement theory, the terms D. Andrich and Ida Marais prefer

R package Dexter: A framework designed for psychometric analysis, particularly focused on educational and psychological testing data such as exams or assessments. Five key features of the Dexter package: **Data** Management, Dichotomous Scoring, Model Support, Item Analysis and Item Moderating

RUMM: Rasch Unidimensional Measurement Models, one of the representative Rasch-based software developed in VB by D. Andrich, B. Sheridan in Australia, and Luo Guanzhong in China (2001)

S

SAT: Scholastic Assessment Test

SCA: Subjective creativity assessments

Scale: The type of information provided by logist about the data in Rasch measurement
Score distribution: The number (or percentage) of test takers at each score level
SEM: Structural Equating Modeling (cf. Mplus)
Separation: An index designed to assess how well a test can distinguish between different levels of the latent trait being measured
Skewness: Peak shape, an index reflecting the symmetry of the data distribution. (Cf. Kurtosis)
Spiraling: The technique often adopted to minimize potential biases and to ensure test fairness. Typically, spiraling involves (1) organizing test items in such a way that different versions of the test contain a similar mix of items, covering all the content areas or skills being assessed; (2) ensuring a balanced distribution of different versions of the same test paper to test takers in adjacent seats who receive different versions from the same test paper in the same testing room. Spiraling is mostly employed in large-scale standardized examinations with high stakes
SPSS: Statistical Package for Social Science
Standard Deviation (SD): The square root of the variance, to standardize the raw scores for the purpose of comparing test scores
Standardized Examination: An examination or a test administered, scored, and interpreted in a standard, predetermined manner; large-scale, norm-referenced tests, using separate and objective test item type that emphasize both ability and speed
Stem: The part of the question presenting the problem to which the test-taker must respond, particularly in the context of multiple-choice questions (MCQ). Stem is the core portion of the question followed typically by four options: Stem, one correct answer with three distractors. In designing a MCQ, the stem is crucial because it must be clearly worded to be unidimentional and local independent
STEM: An acronym that stands for science, technology, engineering, and mathematics. The acronym was introduced in 2001 by scientific administrators at the U.S. National Science Foundation (NSF)
Summation Property: The sum of both item difficulties and person abilities to predict the expected raw score, and in turn, the scores can be interpreted meaningfully across different populations and test forms
SUSS: Singapore University of Social Sciences

T

TAM: Test Analysis Modules, developed by Alexander Robitzsch, Thomas Kiefer and Margaret WU using the R-language
The practice of improving, adjusting test items and the whole structure of a test based on the item analysis
TERA: Taiwan Education Research Association, China
TOEFL: Test of English as Foreign Language

TSM: Technology satisfaction model

Two-Parameter Logistic Model (2PL): This model considers two parameters: Difficulty (b) and discrimination (a), to discriminate between different levels of the latent trait being measured. (Cf. Item Response Theory)

TWCT: Two-way checklist table

Three-Parameter Logistic Model (3PL): This model includes three parameters: Difficulty (b), discrimination (a), and guessing (c), to further discriminate ability with the same performance (scoring) (cf. Item Response Theory)

True score theory: The core of CTT, a framework for understanding the relationship between observed scores on a test and the underlying true scores that the observed scores aim to measure. The key components of true score theory include: True Score, Observed Score, and Error Score. The equation goes as follows: Observed Score = True Score + Error Score (Cf. Classical Test Theory)

U

UAE: United Arab Emirates

UCB: University of California, Berkeley

UCLA: University of California, Los Angeles

UNESCO: United Nations Educational, Scientific and Cultural Organization. A specialized agency of the UN established in 1945 with its mission to promote international collaboration in education, science, and culture.

Unidimensionality: A key assumption in the Rasch measurement model and IRT models, indicating a single latent trait underlying the responses to the items on a test or statements on a questionnaire; a construct or scale that is measured along a single trait or factor. In language testing, a unidimentional scale may measure only one aspect of language proficiency, such as reading comprehension

UiTM: Universiti Teknologi MARA in Malaysia

UNISEL: Universiti Industri Selangor, Malaysia

V

Validity: Also called statistical validity describing how closely scores on a test correlate with performance as measured; test what should be tested in language testing

Variable: A unidimentional attribute with a variety of magnitudes

Variable map: A diagram or a graph outlining or plotting the variables used in a dataset

Variance: Mean square, indicating the dispersion of test data in CTT

Varimax rotation: A statistical technique used in factor analysis to make the factor loading more interpretable by maximizing the variance of the squared loadings for each factor. It helps researchers identify which variables are most strongly associated with each factor so as to enhance the understanding of the underlying dimensions in the dataset

W

WINSTEPS: A kind of statistical software package for data analysis for the social sciences

WSK: Abbreviated from Chinese Characters "Waiyu Shuiping Kaoshi", literally meaning Foreign Language Proficiency Test"

WSU: The World Sports University, Macau, SAR, China

X

X^2 Test: See Chi Square test

Y

Z

Z-score: A statistical measure that indicates the number of SD a data value is away from the mean of a data set

1

The Rasch Model and Related Profiles

Scholars today who understand and are working with the Rasch model as a measurement tool for language testing or other disciplines of social sciences are aware of anecdotal evidence before the Rasch model was generally accepted.[1] It probably started with a big debate between Norman Campbell and social scientists in the 1930s. The issue was whether "psychometrics" was possible. The other stories related to the Rasch model covered in this chapter encompass all that is worth reviewing in the Rasch measurement field. At the same time, basic concepts and terminology related to the Rasch model will be introduced as they help us better understand the rest of the book. The chapter concludes with references from which all the basic concepts and terminology are borrowed for readers to pursue further studies.

1.1 The Early Controversy over "Psychological Measurement in Europe"

In response to the early debate over "Is psychometrics possible?" according to Linacre (2016), stubborn physicists, led by Norman Campbell, said: "Not possible! Measurement requires carefully constructed, or connected calculations (e.g., connecting a stick end to end to measure length, or stacking bricks one on top of the other to measure weight), but you can't connect people's heads to measure people's mental conditions!" Social scientists, led by Stanley Stevens at the time, could not answer such a question well, so they devised their own definition of psychometrics, the new "measurement". "Measurement is the assignment of numbers to measured objects or events according to rules".[2]. Social scientists therefore called whatever numbers they obtained by whatever means "measurement data".

Norman Campbell was a very pragmatic man. Thus, in his book *Physics: The Elements,* he lays out the basic principles of physics. Measurement is at the heart of physics. Norman Campbell's view was clearly the method of "additive measurement". To put it more bluntly, "Measurement means that adding another unit adds the same additional quantity, no matter how much already exists". So the general understanding is that "adding one unit means simply adding 1 to the measurement number, but if you think about "adding the same amount", it gets more complicated. This means defining an operation

DOI: 10.1201/9781003665595-1

that adds a specific constant to the quantity we have. In such cases, for weight, the constant could be one gram or one pound, a practically "stacking" operation. For length, the constant can be one meter or one foot, and the operation is "connecting" (technically a "connecting" calculation). For temperature, the constant can be one degree, which is a "heating" calculation. It follows that Norman Campbell could not imagine how these people could define a unit of "attitude" and then construct a process by which the unit of attitude could be increased by a known quantity (Linacre, 2016). The social scientists of the 1930s failed to face such a problem properly, and so gave up.

1.2 Origin of the Rasch Model

Since then, though, academia has caused some confusion. For social scientists, any number is a metric, as long as you can imagine some rules it follows. For all the others, measurements must conform to strict objective criteria. And only Georg Rasch solved this problem. In 1960, Rasch published his article entitled "A Probabilistic Model of Intelligence and Achievement Tests" which resolved the debate that had originated in the 1930s between physicists and social scientists over the question: "Is psychometric measurement possible?" Rasch himself modestly called it a "measurement model", and since then, this "measurement model" proposed by Rasch himself has been widely used in the fields of educational measurement, sociology, management, psychology, language testing, medicine, and health sciences to deal with variables and has been extended from its initial application to binary data types to multi-valued data types. There is one more little tidbit here. According to Mike Linacre (2016), while Rasch was teaching a course at the University of Chicago, there was only one person who attended and listened to all of his lectures from start to finish, and that person was Benjamin D. Wright (the author of what becomes known in the testing community as Best Test Design). Rasch discovered, invented, and constructed a "measurement model" that essentially overcame Norman Campbell's objections; however, Rasch himself did not claim that. Instead, it was Ben Wright who was spreading Rasch's ideas around the world and made the Rasch model what it is today.

Nowadays, this model is universally known as the "Rasch model" in the academic world, and experts and scholars are convinced that this model will play a great role in the promotion of objective measurement and scientific exploration. In the testing community across the China Mainland, the model is mostly known as "Rasch model", while in the Southeast Asia and China Hong Kong, Macao, and Taiwan, Rasch model is translated as "罗氏" (pronounced as [luoshi]) model. And in today's academic world, there is even an academic society called Pacific-Rim Objective Measurement

Society (PROMS) which is the most authoritative and professional in specializing in the theory of and application to the Rasch model. PROMS organizes the annual international conference on Objective Measurement in the Pacific Rim countries and regions which is a topic of Chapter 6.

Rasch proposed a practical solution based on log-odds conversion. The calculation works well, but social scientists think it too complex (and many still do), while mathematical statisticians take it overly simplistic (and many still see it that way) (Linacre, 2016). For the author, it seems that such an intersection of the arts and sciences and their collision with each other have never been a problem in academia. However, the author would like to remind the readers that for scholars who study language testing or other liberal arts, it has already gone beyond the scope of pure liberal arts to master the Rasch model for their own research. In today's rapid development of computer network technology and in the scientific research in the era of Big Data and Artificial Intelligence, the direction of the said research is certainly more correct, but not necessarily to be more accurate! In this sense, the Rasch model is an effective method for current research in language testing.

1.3 "VIPs" in the Rasch Family

After introducing the origin of the Rasch model, the author thinks it is necessary to briefly introduce the important figures of the Rasch family. These ins and outs are well known in the academic circles of Rasch counterparts and are also shared with scholars and peers in the field of the Rasch model research in other countries and regions.

In the field of the Rasch model research and applications, it is generally accepted that Ben Wright and Panchapakesan are George Rasch's earliest students, albeit originated in Denmark, the Rasch model was actually developed by Ben Wright who brought the model to the United States and taught the ideas at the University of Chicago. Mike Linacre was Ben's research assistant at the time.

David Andrich and Graham Douglas are Ben's early students, followed by Stone and Geoff Masters. Then came Stone and Geoff Masters, who returned respectively to Australia to promote the Rasch model in Sydney, while David Andrich went to Perth to promote the Rasch model. David Andrich has two students who are working hard for PROMS: One is Iris Lee from Singapore, and the other is Luo Guanzhong from China.

Richard Smith, Mark Wilson, William Fisher, and Mary Lunz were all students at about the same time. Among them, Richard Smith was the earlier and produced more articles. One thing worth mentioning here Mark Wilson also has a Chinese student named Huang Xiaoting who is working in Peking University, China. A little later, Jack Stenner and George Engelhard, both of whom played important roles in the PROMS, Jack Stenner's Meta Metrics was a strong sponsor of the PROMS until his retirement in 2016, and George Engelhard is currently Chairman of PROMS (2022–2023) dedicated to his work, keeping up with the times, and steering the PROMS. George is the 2018 Recipient of Benjamin Drake Wright Senior Scholar Award[3] from American Educational Research Association (AERA).

Here in particular, the author would like to make special introduction to three unusual scholars involved in the field of the Rasch model research:

The first one is Prof. Gui Shichun (1930–2017) from China's Guangzhou Institute of Foreign Languages (now Guangdong University of Foreign Studies), my PhD supervisor and a leading scientist in the field of the Rasch model research and application in China. He taught himself the Rasch model but made three outstanding contributions to the successful application of the Rasch model: One is the application of the Rasch model to complete the Ten-year (1990–1999) MET[4] Equating Project sponsored by the Examination Centre under the State Education Commission, MOE, China; the other is the development of GITEST, China's earliest Rasch-based software program, which played a pivotal role in the item calibration and test equating for MET. A third contribution is the practice of Rasch-based MET test equating earned the project "China's Equating Model". That's Good old days for Chinese psychometricians!

The second is Prof. Magdalena M. C. Mok from the Hong Kong Institute of Education (now the Hong Kong University of Education). Professor Mok jokingly calls herself the "dwarf" among the Rasch senior counterparts. In terms of seniority, she studied under Ben and has made a lot of achievements in the Rasch model research, but she has always been a quiet and humble person. The greatest contribution made by Prof. Mok to the research and development of the Rasch model in China is that together with Mr. Sammy in 2012, she introduced the PROMS to China mainland,[5] which led to many Chinese scholars in the research field to go beyond Asia to reach the international level today.

The third one is Prof. Trevor G. Bond, professor at James Cook University in Australia with great international influence. Bond was never enrolled formally as a student of the Rasch model research. He was mentored by Ben Wright and Mike Linacre in his early days, but it was his persistence in understanding the Rasch model that gave him his current position. Although Bond has written the most influential Rasch monographs, conducted workshops in Southeast Asia, and taught the Rasch model, in his own words, "I have never considered myself an expert of Rasch measurement" like many others do. However, in fact, he has, and even more! Bond's greatest contribution to the research and application of the Rasch model was founding the "Pacific Rim Objective Measurement Symposium (PROMS)" in 2005 and serving as the president of PROMS for almost a decade. At that position, he organized the annual PROMS conferences which greatly promoted and popularized the Rasch model in the Pacific Rim regions, countries, and even the world. In 2024, the Rasch SIG is thrilled to announce Dr. Trevor Bond as the winner of the Rasch SIG's bi-annual Benjamin Drake Wright Senior Scholar Award. To collect his award, Bond will present an invited address at the Rasch Measurement SIG Business Meeting session at AERA in Denver, Colorado in April 2025.

1.4 The Debate over "Standardized Examinations" in China

If the early "psychometric debate" gave birth to the Rasch model, the debate over "standardized test" in China has accelerated the successful application of the Rasch model to item analysis and test equating. The author here briefly reviews the misunderstandings about standardized tests (in fact, about the Gaokao) and about the process of test item evaluation yielding from non-professionals both in and outside schools as well as from parents in particular since 1985 when the standardized tests were first implemented in China Mainland. To cut the story short, standardized examinations were simply thought of as multiple-choice questions; test items were conflated with test papers, and so on. Little was known about the test theory inherent in the practice of standardized tests. Of course, such misconceptions were quickly cleared up[6] (for details, see Section 2.2 of Chapter 2).

Now it seems that it is imperative to promote the basic knowledge of standardized tests at all times in China, because any large-scale test is guided by certain theories; for example, how to calibrate or assess the test items used in standardized tests? According to Prof. Gui Shichun (1995), this can be done from both the internal and external perspectives of the test: Internal assessment is related to reliability, while external assessment is related to validity. How to ensure the fairness of the MET scores for thousands of families? These can be achieved through test equating via the Rasch model. It is also for this reason that the Examination Centre under the Ministry of Education of China had launched the Ten-Year (1990–1999) MET Equating Project, which did solve these problems well. In addition to the Rasch model, language testing now also integrated with modern cognitive science, computer science, and artificial intelligence has come of age, showing the development of a discipline, thus marking the end of an era, namely: the end of the era of paper-and-pencil testing; and the arrival of a new era, namely: the arrival of the era of computerized cognitive testing, online and network testing! Therefore, language testing is no longer about scores in the descriptive statistics, but also about item analyses, test equating, and item banking. The application of multimedia technology, cognitive science, and artificial intelligence to language testing has made it possible to start exploring the internal cognitive and psychological processes of human beings through the observable scores.

1.5 China's Ten-Year MET Equating Project for the Entrance of Higher Education

In order to further dispel misconceptions about the Gaokao and to resolve the problem of scores equating over the past years, the Chinese Ministry of Education (then the Examination Centre of the State Education Commission) implemented the "Ten-year MET Equating Project" (Gui et al., 1993). That is, following Wright and Stone (1979), the project team adjusted the scores of the national entrance exams for higher education over the years by using anchor items. In layman's terms, the scores obtained from a particular year's MET paper were used as a basal, and then calculate the difference between the current year's scores and the basal year's scores each year so as to re-scale the scores to be consistent with the basal year's scores. In this way, the scores obtained from two years are comparable on the same scale. According to the Rasch model, the difficulty of test items is relatively independent, and the MET project team can use sample data from fewer observation points (actually middle schools) to calculate the overall level of different abilities.

The basic model of MET equating goes thus: MET1988 was taken as the basal test, and a balanced test paper was designed exactly the same in format as the MET paper. The reason why we call it "balanced test paper" is that such a balanced test paper contains **linking items;** to collect the sample data, take the balanced test paper to the candidates at the observation points three days before the annual MET was administered, and then the sampled test takers take the current year's MET papers three days later together with all the candidates across the country. In doing so, we have managed to obtain both scores of the balance test paper and of the current year's MET test performed by the same sampled candidates. With these two sets of sampled scores, we can adjust the difficulty of each year's MET so that the test takers' raw scores are equated to those of MET 1988.

The steps for conducting MET equating were carried out in the following three ways: First, the balanced test paper used at the observed middle schools consists of 85 multiple-choice questions, of which 35 ones were used as linking items. These 35 linking items are actually the anchor items that link the two examinations (i.e., the basal and the current). In order to ensure reliability, a chi-square test was carried out to eliminate the items that do not meet the requirements.

Second, MET1988 paper was used as the basal test calibration, connected via linking items, using GITEST to first obtain the difficulty values of both linking items and the basal test items, and then to obtain the difficulty values of both linking items and the current year's MET test items Finally, the difficulty values of both basal test and the current year's MET are adjusted or rescaled based on the difference between the two means of the same linking items.

In the final step, based on adjusted difficulty parameters, the candidate's ability can be estimated or re-scaled, thus comparable on the same scale. In this way, the difficulty values of all the MET test papers over the past years are equated and comparable on the same scale. To make the whole society understand more about the significance of the MET equating practice, the equating group also converted the difficulty parameters into concepts and ability estimators into percentage, so as to popularize the understanding of the post-equating results of MET. Taking the example of Guangdong Province in China, the average score of the candidates in the whole province at that time was 44.4, indicating that the English proficiency of the candidates in 1989 had improved by 4.4 points compared with that of the candidates in 1988; the average value in 1990 was 53.7, which was another increase of about 9 points compared with that of 1989, 50.0 in 1991, and 54.2 in 1992. If the scores were reported on the basis of the raw scores without any adjustments, then the mean would have been, respectively, 44.8 in 1988, 48.0 in 1989, 42.0 in 1990, 43.2 in 1991, and 44.4 in 1992. The level of competence of candidates in 1990 would otherwise be considered the lowest. This shows that the test equating is compulsory and that the results are in line with the actual situation in China.

In short, after entering the 21st century, education, examination, and scientific research have stepped into an era of high technology incorporated with artificial intelligence, and everything requires scientific and systematic operation in nature. As a tool and means of educational measurement, large-scale yet high-stakes entrance examinations like MET should be even more so. It is not only related to the reliability of candidates' scores and the fairness of the examinations, but also directly related to the evaluation and promotion of the entire education system. In this sense, the successful implementation of the Ten-year MET Equating Project by applying the Rasch model in China reflects the on-going requirements of the times.

1.6 The Belated "Rasch Wars"

The "Rasch Wars"[7] refers to a period of intense debate and discussion within the field of language testing, primarily in the late 20th century and particularly concerning the application and implications of the Rasch model in educational and psychological measurement. They involved academics, psychometricians, and language testing professionals who were divided over the merits and limitations of the Rasch model compared to other statistical models and methods used in educational assessment and language testing. It is known to all, the Rasch model, developed by Georg Rasch (1901–1981), is a probabilistic model used for analyzing data from assessments to measure variables such as abilities, attitudes, or personality traits. Therefore, the "wars" aspect of this term highlights the strong disagreements and differing perspectives among scholars and practitioners regarding the appropriateness and utility of the Rasch model compared to other statistical models such as classical testing theory (CTT) or item response theory (IRT) models.

1.6.1 Pros and Cons in the "Rasch Wars"

Since the debate has already ended and the supporters have won a complete victory, the author no longer needs to name who is in favor and who is against. To be in general, in the context of the "Rasch Wars", the representative of the pros of the Rasch model is associated with Benjamin D. Wright, the prominent advocate for the Rasch model and played a significant role in promoting its use in educational and psychological measurement. Wright argued for the advantages of the Rasch model, particularly focusing on invariant measurement (e.g., Engelhard, 2013; Engelhard and Wang, 2024) and its ability to provide meaningful, interval-level data from ordinal raw scores. Wright's work, along with that of his colleagues and followers, emphasized the model's potential to enhance the fairness and precision of assessments, contributing significantly to the debates and discussions that characterized the Rasch Wars. While these debates highlighted the strengths and weaknesses of the Rasch model and led to broader discussions about measurement practices in general, the author just narrowed down the debate to some of the pros and cons typically associated with language testing in specific.

Those who advocated for the Rasch model appreciated its strong theoretical foundations, its focus on invariant measurement, and its ability to provide interval-level measurement from ordinal data mainly covering three aspects, while opponents were concerned about its assumptions and limitations, such as the requirement for unidimensionality and the model's fit to data mainly including three points. Table 1.1 summarized both ideas.

TABLE 1.1

Pros and Cons of the Rasch Model for Language Testing

Pros for the Rasch Model for Language Testing	Cons against the Rasch Model for Language Testing
Unidimensionality: The proponents assume that a single latent trait underlies the responses, which simplifies the interpretation of test scores.	**Assumptions:** Critics argue that the assumptions of unidimensionality and local independence may not hold in practice, limiting the model's applicability.
Invariance: The proponents advocate the model provides parameter estimates that are invariant across different samples and test forms, assuming the model fits the data well.	**Complexity and Limitations:** Only some argue that the Rasch model's constraints, such as requiring all items to have the same discrimination parameter, can be overly restrictive and may not capture the complexity of human responses as well as other IRT models.
Interval-level Measurement: The proponents address that the model transforms ordinal raw scores (via different Rasch-based software programs) into interval-level measures, which are more suitable for statistical analysis.	**Empirical Fit:** There are concerns about the model's fit to real-world data, as it may not always accurately represent the underlying constructs.

1.6.2 The Features of the "Rasch Wars"

Overall, the Rasch Wars contributed to the evolution of psychometric practices by encouraging rigorous examination and discussion of measurement models, though they also highlighted the challenges of integrating complex statistical methods into practical testing scenarios. In fact, this debate has been particularly relevant in the language testing community, where accurate measurement of both language proficiency and other constructs is crucial. The debates were characterized by discussions on methodological rigor, philosophical differences regarding measurement, and practical implications for test development and analysis, influencing how language assessments are developed and interpreted.

It is not the intention of the present author to elaborate here the main points of the debates, as they largely revolved around two aspects: One was the debate about the appropriate applicability of the Rasch model to language test data (especially dichotomously scored data); the other was the tiresome debate over or the cliché of the Rasch model versus the one-parameter model of IRT. These discussions took place seven years after Bond founded the PROMS in Malaysia (see Chapter 6), and more than a decade years after the successful implementation of China's Gao-Kao equating project. During those quiet days, the use of the Rasch model has been prevailing in the field of language testing. Here, the author's reference to the discussion about the Rasch model is another way to further illustrate the fact that the appropriateness of research and application of the Rasch model in the field of language testing has long been indisputable. No wonder in the present author's own experience, I never heard Bond say anything about "Rasch Wars", though in his keynote speech at PROMS2024, Kuala Lumpur, Malaysia, Bond did use "Coup d'état", but not referring to that discussion. Neither did Prof. Gui mention anything about "this war". The author would like to quote verbum pro verbo Peter Tymms (2021) in the foreword of the fourth edition of Bond's most influential monograph to justify the status and role of the Rasch model in research and application in the humanities and social sciences[8]:

> When Georg Rasch wrote his seminal book in 1960 he started a revolution which was to change the way in which measurement in the social sciences was conceived. However, as with all revolutions the message takes time to spread; starting first within original discipline, Education, moved to other fields broadening its base all the time. For this to happen those who saw the potential of Rasch measurement needed to promulgate the ideas and the most notable of these was Ben Wright from the University of Chicago, who was proved to be an effective communicator and innovative developer. He became the PhD supervisor of a series of bright students who, in their turn, developed the model further and took the ideas to others. Some of Wright's inspirational lectures can be seen on YouTube. As Rasch message spread, there were more and more converts

> who saw the light, so to speak, and They could take on an aura of religious fervor so much so that, on occasions, the SIG meeting at the AERA annual conference felt like a prayer meeting including a discussion as to how to convert the ignorant and unbelievers.
>
> **(Peter Tymmms, 2021, p. xxx)**

Peter then expressed quite bluntly that the 4th edition of Bond's works: *Applying the Rasch Model: Fundamental Measurement in Human Sciences* is an authoritative and well-argued monograph that signals the dissemination and development of the latest results in this important field of measurement.

1.7 Basic Concepts of the Rasch Model

The basic concepts regarding the Rasch model in this section are either borrowed or selected from the references at the end of this chapter, and primarily intended for students of liberal arts, social sciences. For more detailed basic concepts and principles of the Rash model, the author suggests the perusal of the books listed at the end of this chapter. A good understanding of these basic concepts helps guide us in using the Rasch model and any Rasch-based software properly.

1.7.1 Latent Traits

Let's start from the concept of "latent traits" to understand the ideas of the Rasch model analysis. In the social sciences, "latent traits" are referred to as attributes or (unobservable) properties that cannot be measured directly. In Rasch measurement contexts, where scoring is associated with such latent traits, the idea is subject to various attributes or properties such as language grammar or mathematics. To quote Andrich (pp. 3–4, 2019), it is such properties of entities that are to be measured, not the entities themselves. A researcher does not measure a person but a psychological attribute (property) of the person such as neuroticism or intelligence.

1.7.2 Rasch's Dichotomous Scoring Model

Rasch proposed several mathematical models in his book entitled Probabilistic Models of Intelligence and Exit Tests (Chicago, 1980). His model regarding test takers' responses to dichotomous ratings (right/wrong, yes/no, present/absent) of test items is well known in the field as the "Rasch model". Dichotomously scored data is the very type of data that is used today. This is the starting point for this book and the basic approach we use to analyze test data. Rasch originally used Greek letters to represent his model in multiplicative form. Later authors have used Greek letters to represent the same concept in exponential form. But per Mike Linacre (2016), some scholars argued that both forms obscure the underlying additive process; therefore, the more familiar Latin (Roman) alphabet is adopted to represent the model in logarithmic form as follows:

$$P_i = \frac{\exp(\emptyset - b_i)}{1 + \exp(\emptyset - b_i)} \tag{1.1}$$

This is the original Rasch model. The binary-valued scoring model specifies that a test taker n with an ability of B_n has a probability P of succeeding on an item i with a difficulty of D_i. According to Rasch, "success" means "demonstrating more of the latent variables that we would expect. "Failure" means "exhibiting fewer of the variables we expect". In the case of language tests, it refers to that "getting a correct answer" means "test takers showing more of the potential variables we predicted in the test". An "incorrect answer" means that "the test takers demonstrating fewer of the variables we predicted", so we must rate observed scores according to this assumption, regardless of the values assigned to the observation during the data collection process.

Therefore, we have:

$$\log_e \left\langle \frac{P_{ni}}{1 - P_{ni}} \right\rangle = B_n - D_i \tag{1.2}$$

Formula (1.2) states that "The logarithmic chance of candidate n successfully answering item i" is equal to "test taker n's ability minus difficulty of item i".

1.7.3 $\log_e$ Concepts

The reference to $\log_e$ in the Rasch model is the notion of a "natural" logarithm. Log is abbreviated from logarithm.[9] The natural logarithm is one of the most useful functions in mathematics, with applications throughout the social sciences.

1.7.4 Probability

Rasch refers to probability in his model in terms of frequency (Mike Linacre, 2016), i.e., the probability of an event is the proportion of times the event would occur if we could repeat the operation many times. According to Rasch, the probability is always between 0 (never happens) and 1 (always happens). In language testing, P is used to refer to the probability of a correct answer, and 1-P is the probability of a wrong answer. Since a right or wrong answer must always occur, when we add up their probabilities, the sum would be 1; i.e.,

$$\begin{aligned}&(\text{Probability of occurrence})+(\text{Probability of non}-\text{occurrence})\\&\qquad=(\text{P})+(1-\text{P})=1.\end{aligned} \tag{1.3}$$

What if the probability of getting it right is the same as the probability of getting it wrong? Then both probabilities are 0.5. The odds of getting it right are 0.5/0.5 = 1, i.e., $P_{ni1}+P_{ni0}=1$, and the logarithm of the odds is log (1) = 0. So ability and difficulty are consistent. By Rasch's principle, ability and difficulty are on the same scale, which is exactly what we would expect. In language testing, when a test taker is presented with an item of exactly the same difficulty as his or her ability, the test taker is often unable to predict whether he or she will get it right or wrong. Prediction at this point is like flipping a coin: Heads or Tails?

To help us understand the Rasch model more specifically in the context of language testing, let's explicitly stipulate that the practice of Rasch model of dichotomous scoring is done by scoring one correct answer with one point and one wrong answer with zero point. This is what is meant by dichotomous scoring. Given this, the Rasch dichotomous scoring model specifies the probability P_{nil} that a test taker with an ability of B_n scores a 1 on item i with a difficulty of D_i and P_{ni0} as the probability of scoring 0. It can be further clarified here that the so-called proficiency is a concept that we seek among test takers in language tests. In the case of other research disciplines, Your Excellency may use other words to indicate your field of study, e.g., "motivation", "comprehension", "proficiency", etc. "Difficulty" is what we specify in the test items.

B_n and D_i are logits of distance along the latent variable relative to the local starting point. A "latent variable" is something we can have more or less of, but we cannot measure it directly. It is a variable like comprehension or writing quality. We conceptualize it as a straight line with equally spaced output units. The line is infinite. We can always imagine that something (or someone) has more attributes than anything (or anyone) we have encountered so far, or that something (or someone) has fewer attributes. We conceptualize each observation in the data to represent the potential variable of "less" or "more".

1.7.5 Qualitatively Ordered Data

"Qualitative-ordered data" refers to what we often say; e.g., "good, better, best". In language testing, a "correct answer" means "more of what we are looking for". A "wrong answer" means "less of what we are looking for".

The difference between "right" and "wrong" answers is qualitative. The qualitative ordering of these different qualities is shown by rating them "1" and "0". The standard Rasch dichotomous rating goes thus:

"1" means "more potential variables".
"0" means "fewer potential variables".

These qualitatively ordered data formats are ubiquitous in language testing or in academic research in general.

1.7.6 Concept of Logits

"Logits" are "logarithmic odds units". Refer to the above (Formula 1.2):

$$\log_e\left\langle\frac{P_{ni}}{1-P_{ni}}\right\rangle = B_n - D_i$$

On the left is "$\log_e$" for "natural logarithm".

P_{ni1} dividing the logarithm of P_{ni0} by the logarithm is the ratio of the two probabilities. The ratio of the probabilities is called the "odds". Therefore, on the left side of the equation, we have log (odds) units, and these provide the units on the right side of the equation, so B_n and D_i are measured in "logarithmic odds units", i.e., logits, pronounced [lôdjits]. Typically, logit values are linearly adjusted (multiplied by a constant), then another constant is added to produce numbers that are easier for most people to understand. It's like converting from 38° Fahrenheit to 3.33° Celsius on a thermometer. The meaning of the indicated temperature remains the same, but the number is different.

1.7.7 Local Starting Point

In the context of the Rasch model measurement, "local origin" means the starting point of our measurement. In language testing, 0 is located at the center point of the item difficulty/subject ability scale, which can be interpreted as medium difficulty/ability, usually represented on the measurement scale as 3, 2, 1, 0, +1, +2, +3.

The term "local starting point" can be subject to different interpretations. In the context of fitting model to data, local starting point may refer to initial estimates used for the iterative process of parameter estimation wherein the specific choice of starting values can be taken as "local starting point". In general, these starting values are set mostly based on heuristic methods or prior knowledge. Specifically, all item difficulties may be initialized to zero, or person abilities to be set to the mean of the data to start the estimation process. The choice of these starting values may affect the convergence and speed of the estimation procedures. If specific software is used for Rasch analysis, the software may have particular ways of handling starting values.

In the case of BILOG or PARSCALE, the user should specify the KEY WORD CRIt=? (the default of which is 0.001) in the command file to set the appropriate starting point to begin iteration. For more details, the readers may consult the software user guide (3.8).

1.7.8 Calibration

In the Rasch model study, technically, calibration refers to the process of converting raw scores to logits on a Rasch measurement scale. There are five aspects involved in the calibration process such as item difficulty estimation, person ability estimation, fit analysis, scale development, and iterative process. Item difficulty is a parameter indicating how challenging an item is for test takers. Alongside item difficulty, calibration also estimates the ability level of each test taker. As we all know, the Rasch model posits that test takers' ability and test item difficulty are comparable on the same scale, thus ensuring meaningful comparisons between different test takers and test items. During calibration, the fit of the data to the model is assessed. This is actually checking whether the raw data match the expectations of the model. Goodness of fit (see Section 1.7.9) indicates that the model is appropriate for the data and that the items are working as expected. In what follows, calibration generates a scale which goes invariant (see Section 1.7.14) across different groups of test takers. This shows that the measurement properties of the items remain consistent regardless of the sample, thus achieving fairness in comparisons. Finally, calibration is basically an iterative process (see Section 1.7.10). Initial estimates are getting refined through repeated analysis of cycles so as to come to better convergence, showing good fit of the model to the data.

Calibration provides researchers and practitioners with reliable and valid measures of the latent trait being assessed. Given correct data matrix, using any one of the Rasch-based software (see Section 3.8, Chapter 3) can fulfil the task.

1.7.9 Residual

Literally, "residual" means something that remains after the main part has been removed. In the context of the Rasch model study, a residual (R) highlights the discrepancy between an observed value (y) and the value predicted by the model ($\hat{y}$). As Rasch probabilities are expressed as fractions and actual responses are recorded as whole numbers, every item/person interaction in a data matrix will yield some residual value; therefore, we have $\mathbf{R} = \mathbf{y} - \hat{\mathbf{y}}$, where large residuals indicate poor fitness and contribute to misfit or unfit. From another perspective, large residuals would prevent iterations from coming to convergence. In this sense, a good analysis of residuals helps researchers and practitioners determine if the data fit the model or if any modifications are needed.

1.7.10 Goodness of Fit

By goodness of fit is referred to as an important indicator regarding the degree of matching between the actually observed data and the predicted results generated by any Rasch-based software.

According to Frank B. Baker and Seock-Ho Kim (2004), one of the criteria for evaluating any model is how well it replicates the observed data. The closer the fit between what the model predicts and what is observed, the better the model. Goodness of fit is the core issue of the Rasch model (Bond, 2021) in language testing. It can also be said that such item fitness is an indicator used to measure the explanatory power of the Rasch model on observed data. Smith (1991, 1992, 2000), Smith and Miao (1994), Wright and Master (1982), Wright and Stone (1979) (as cited in Bond, 2021) all agreed that because visual inspection of a data matrix typically is not practice for assessing the impact of individual or persons. Rasch analysis provides fit statistics designed to aid the investigator in making a number of interrelated decisions about the data.

In practical application, the smaller the fitness index is obtained, or the closer it is to 1, the better fitting effect it will indicate.

1.7.11 Iteration and Convergence

Iteration and convergence are a pair of terminology commonly encountered when applying Rasch models. **Iteration** refers to the process of starting from an initial estimate, gradually approaching the true solution through the iterative process, and then gradually approaching a fixed point, which is called **convergence**. To determine whether an iteration converges, it is usually necessary to check the convergence domain and convergence speed of the iteration formula. In fact, convergence is determined by the differences between the iteration cycles. If the difference is less than or reaches a pre-determined value, it can be considered that the iteration has come to convergence. The parameters thus obtained through iterative convergence are generally considered to be relatively reliable. Users of software such as BILOG or PARSCALE know that there is a KEY WORD (CRT =?) in the command file, which is used to set the pre-determined value for iterative convergence.

1.7.12 Concepts of the Logistic Ogiv

Reasoning on Rasch's account, if a test taker is most certain to get a correct answer, the probability that the test taker will get it right is 1, and the probability that he or she will get it wrong is 0. The test taker's ability to logarithmically match the number will be positively infinitesimal with respect to any item of finite difficulty.

If the answer is sure to be wrong, then the probability of getting it right is 0 and the probability of getting it wrong is 1. Then the test taker's logarithmic ability, relative to any item of finite difficulty, will be negatively infinite.

Suppose a test taker gets 3 items correct out of 4 answers. Then the probability of the test taker answering correctly is:

$$P = 3/4 = 0.75$$

The probability of a wrong answer is:

$$1 - p = 1/4 = 0.25$$

Then its chances of a correct answer are:

$$P/(1 - P) = 0.75/0.25 = 3$$

This means that the test taker's ability is $\log_e(3) = 1.1$ log units, implying that the ability value is greater than the difficulty value of the test item answered correctly by 1.1 log units. We expand this idea a bit to obtain a relationship between ability (relative to the test item) and the probability of answering it correctly. This relationship is known as the "logistic ogiv" in the Rasch model studies, and is shown in the following diagram: logarithmic difference $(B_n - D_i)$

In Figure 1.1, where the ability of 0 (shown by the dotted line) corresponds to a probability of 0.5 correct answer. An ability advantage of 1.1 log units corresponds to a probability of 0.75 (shown by the solid arrow). The logistic stripes are "monotonically increasing" curves. The probability of getting correct answers always increases as the measurement difference increases.

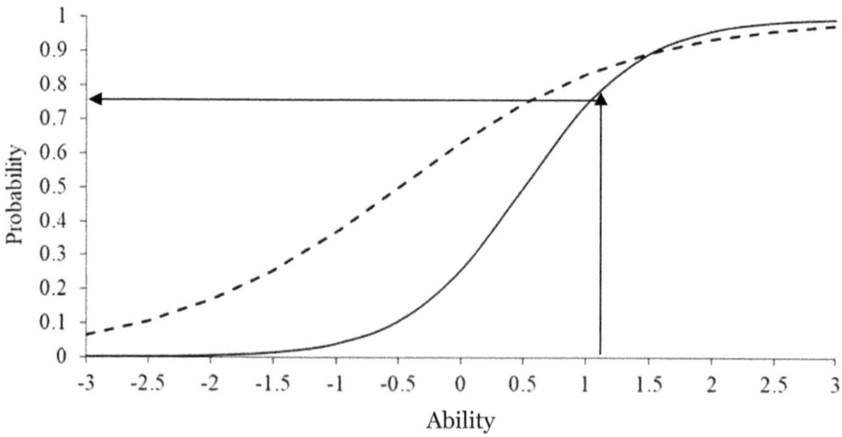

FIGURE 1.1
The ideas of the Logistic Ogiv in the Rasch model

1.7.13 Unidimensionality

In the Rasch model study, unidimentionality is an important hypothesis which refers to a single latent trait or ability underlying the responses to the items on a test or statements on a questionnaire. In psychology and social sciences, the concept is understood as a construct or scale that is measured along a single trait or factor. In language testing, for example, a unidimentional scale may measure only one aspect of a test taker's language proficiency, such as reading comprehension. According to Bond et al. (2021), unidimentionality is a basic concept in Rasch measurement reflecting the Rasch model's focus on the process of fundamental measurement. It is a prerequisite that the data fit the model sufficiently in order to achieve invariant interval-level measurement. Keeves (1997) (as cited in Bond et al., 2021) reminded us that unidimentionality does not imply that performance on items is due to a single psychological process. In fact, a variety of psychological processes are involved in responding to a set of test items. However, as long as they function in unison—that is, performance on each test item is affected by the same process and in the same form—unidimentionality will hold.

1.7.14 Item and Person Parameters

In Rasch model study, item and person parameters are two important parameters. In language testing, they refer to item difficulty of each test item and subject ability estimate of each test taker. With this, the probability of a correct response (of dichotomous value) is modeled as a function of the difference between the subject ability and the item's difficulty. Both item and person parameters of logit values are comparable on the same scale. If WINSTEPS or Rasch-GZ is used, both items and persons are located on the same (Wright) map, wherein the relationship between the distributions of item and subject measures plotted along a vertical logit scale. The highest values are located at the top of the map and the lowest values are located at the bottom, which looks very intuitive (for details, see Chapters 4 and 5) .

1.7.15 Invariance

Invariant is one of the key properties inherent in the Rasch Model. By invariance is referred to, under certain conditions wherein the model fits the data well, as the parameters of item difficulty are independent of the sample of test takers, and the estimates of subject ability are independent of the sample of items.

Thurstone and Rasch (as cited in Andrich, 2019), the requirements of invariance are not descriptions of any data set. They are requirements that data need to meet if they are to be used in measurement. According to Engelhard (2013) and Engelhard and Wang (2024), the goal of measurement is to develop variable maps that can be used to represent the locations of both person and

TABLE 1.2

The Five Essential Requirements of Invariant Measurement Proposed by Engelhard and Wang

Person Measurement:

1. The measurement of persons must be independent of the particular items that happen to be used for the measuring: Item-invariant measurement of persons.
2. A more able person must always have a better chance of success on an item than a less able person: non-crossing person response functions.

Item Calibration:

3. The calibration of the items must be independent of the particular persons used for calibration: Person-invariant calibration of test items.
4. Any person must have a better chance of success on an easy item than on a more difficult item: non-crossing item response functions.

Variable Map:

5. Items and person must be simultaneously located on a single underlying latent variable.

items on a latent variable of substantive and theoretical interest. In order to develop useful scales for measurement, there are five essential requirements of invariant measurement which are listed in Table 1.2.

These five requirements are based on the philosophical and theoretical framework of Rasch measurement models. They reflect an ideal-type model for measurement in the human sciences. (For details, see pp. 13–14, Engelhard and Wang (2024).

1.7.16 Additivity

In Rasch model research, additivity is defined as the essential property of the model, allowing for the measurement of potential features such as ability or motivation on a linear scale. In language testing, when the model is said to be additive, it is in the sense that the difference between subject ability and item difficulty directly determines the probability of a correct answer. In the context of the Rasch study, additivity is understood mainly in five aspects such as invariance, linear measurement, log-odds units, separation of person and item parameters, and summation property.

(1) **Invariance**
See the proceeding Section 1.7.12.

(2) **Linear measurement**
In Rasch model study, linear measurement refers to raw scores being converted into measures on a linear scale on which the difference between any two points represents a consistent difference corresponding to the latent trait being measured, regardless of where (on the scale) the points are located.

(3) **Log-Odds**
Bond (2021) defined log odds as "a logarithmic transformation of the odds to probability of success". The probability of a correct response in terms of log-odds is additive, i.e., the difference in log-odds between any two items or persons can be added or subtracted. In the Rasch model study, this is a very important concept, showing a straightforward interpretation of differences in subject ability or item difficulty (also refer to Section 1.7.5).

(4) **Separation of person and item parameters**
We should understand the idea of separation first. In Rasch analysis, the separation refers to an index designed to assess how well a test can distinguish between different levels of the latent trait being measured. The index usually falls into two types: Person separation index (PSI) and item separation index (ISI). A high PSI (typically > 2.0) suggests that the test is better at discriminating between high and low ability of test takers. On the other hand, a higher ISI suggests that the test items are spreading out across a range of difficulties, allowing for a more detailed analysis of the trait being measured. In contrast, a low ISI indicates that the test may not be reliably displaying between different levels of test item difficulties which could be centralized within a narrow range on a linear scale.

Apart from this, both indexes are related to reliability measures, and they provide a specific perspective on the nuanced measurement of the Rasch model. Both are good estimates of how consistently the test can separate test takers or test items.

Additivity accepts both PSI and ISI. This means the estimated ability of a person does not depend on the specific test items they answered, and the difficulty of a test item does not depend on the specific sample of people who attempted it.

(5) **Summation property**
In the Rasch model study, summation property is interpreted as the sum of both item difficulties and person abilities that can be used to predict the expected raw score, and in turn, the scores can be interpreted meaningfully across different populations and test forms.

1.7.17 Common Item Linking (CIL) and Common Subject Linking (CSL)

In the Rasch model study, CIL and CSL are two crucial concepts used for parallel test equating. By CIL is referred to as such a practice that the same set of well-calibrated test items is put in two separate test papers: one as basal test and the other to be equated. And the tests are administered to two different groups of homogeneous samples at different times. This is the parallel test equating, and the data thus collected, via calibration, produce

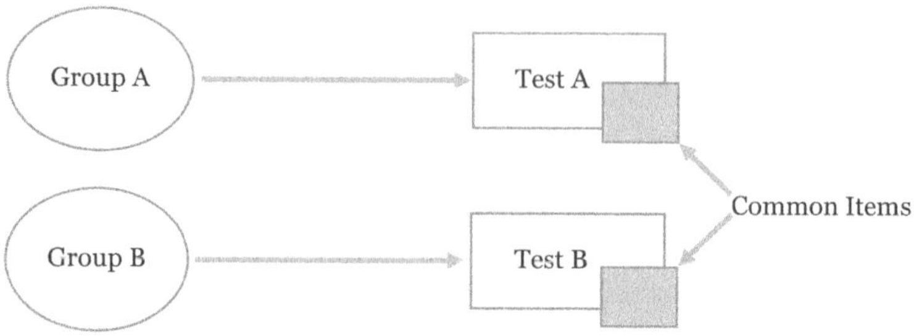

FIGURE 1.2
The ideas regarding common item linking (CIL)

the item difficulties and subject abilities which are comparable on the same scale (for details, see Chapter 3). The ideas are best illustrated in Figures 1.2 and 1.3.

Different from CIL, CSL is referred to as such a practice that two different test papers are administered to the same group of sampled test takers. This is the typical test equating via common subjects. As the data are obtained from the same group of people, the item difficulties and ability estimate via calibration are comparable on the same scale. The ideas are best illustrated in Figure 1.4.

1.7.18 Anchor

In the Rasch model study, anchor is a different term used to refer to the same idea of "common items". Therefore, to great extent, within the context of language testing, "common items" and "anchored items" can be used alternatively to mean the same thing (for details, see Chapter 3).

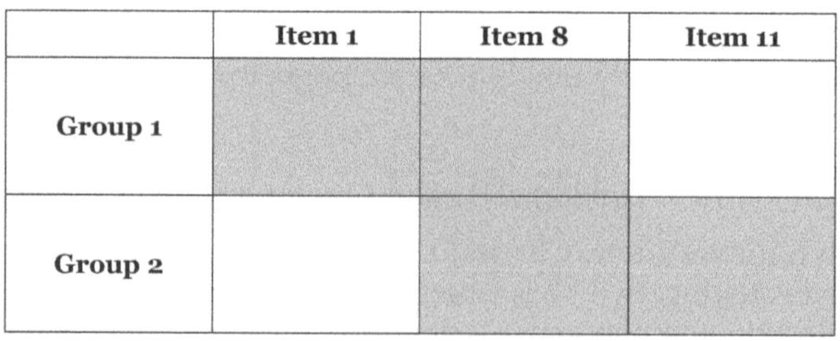

FIGURE 1.3
The ideas about linking design for three items with one item common to two groups (Andrich, 2019, p. 138)

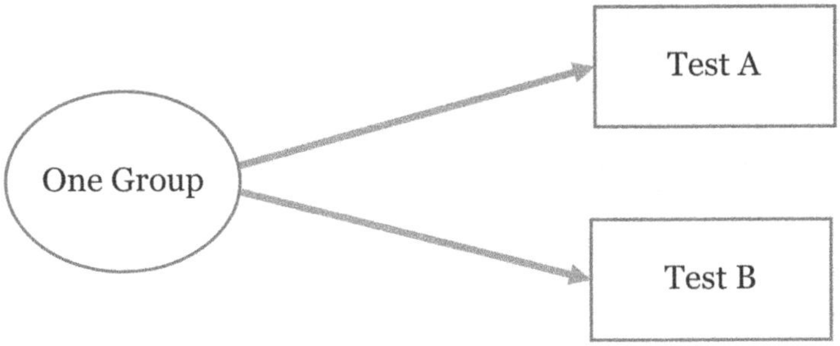

FIGURE 1.4
The ideas regarding common subject linking (CSL)

1.8 Rasch Model and Classical Testing Theory

As we all know, CTT began in the late 19th and early 20th centuries and has been widely used in the fields of educational measurement, language testing and so on. The assumption of true scores in CTT is its core element. According to CTT (Lord and Novick, 1992), the observed score is equal to the sum of the true score and the measurement error, i.e.,

$$O = T + E \tag{1.4}$$

where

O denotes Observed scores, the raw scores actually received by the test takers;

T refers to True score, an abstract assumption about the true abilities or traits of the test takers; and

E indicates random measurement error (Error).

The theory of true scores is made up of these three elements. From this, we can deduce that the correlation between the true fraction and the random error fraction is zero ($r = 0$), which can also be interpreted to mean that (T) and (E) are independent of each other; the other point is that in actual measurements, errors are unavoidable, and that (O) ≈ (T) or = (T) only if (E) ≈ 0, or = 0. In other words, the observed fraction is infinitely close to the true fraction only when the random measurement error is infinitely close to ∞ 0. Among the classical measurement theories, the theory of true scores is considered to be the earliest measurement theory to be mathematically formalized. In addition to this, the CTT also proposes the concepts of Reliability and Validity, etc. The CTT, despite its limitations, contributes a lot to the item analysis in language testing (see Chapter 2 for more details).

Andrich and Ida Marais (2019, pp. 149–150) once summarized some comparisons and contrasts between CTT and what they call RMT, i.e., Rasch Measurement Theory.[10] In their opinion, the motivation of the theories and models of CTT and RMT appears so different that they consider the two theories are incompatible. However, instead of dichotomously accepting one or rejecting the other, they contribute the differences regarding assumptions and models as an elaboration of RMT to CTT. In what follows, the author justifies the position by addressing some differences between the two from the perspective of language testing.

The Rasch model differs from CTT in three aspects. The first is unidimentional assumption; for example, it is assumed that each question or test item in a test measures a latent trait together and that this single latent trait is inherent in all the test items. A test taker's performance on a test can only be explained by one latent trait (e.g., grammatical ability). To use Bond's words (p. 34, 2021), such a focus on one attribute or dimension at a time is referred to as **unidimensionality**. Next is the assumption of local independence. That is, a test taker's response to a test item is influenced only by his or her own level of ability, and the test items are so designed that a test taker's correct or incorrect answer to one test item does not give any clue to the next test item to be answered correctly or incorrectly. This is what the ability (trait) is referred to as in the Rasch model; the third point is that the difficulty values of the test items generated by the Rasch model and the test taker's ability estimates can be placed on the same scale for comparison.

Because of this, the Rasch model is superior to classical measurement theory in three ways: First, its estimation of item parameters is stable and unaffected by the sample of test takers, and iteration using Rasch-based software is easy to converge, thus the parameter estimation is much more accurate. Second, the Rasch model can even give a (individually varying) measurement error for each test taker and accurately estimate test taker's ability. Third, perhaps the greatest advantage of Rasch model over CTT is found in solution to the problem of parallel test equating, which not only equates the total score (i.e., ability value) of the test takers' performance but also equates the difficulty parameter of the test items. This goes beyond the reach of CTT.

At present in China, the CTT is still the main measurement method used by schools and examination authorities. Its shortcoming is obvious: the estimation of the test item difficulty is dependent on the ability level of the test takers, which in turn constrains the difficulty of the test items. This can also be interpreted from another perspective, that is, the scores of different test papers taken by different groups of test takers are by no means comparable in terms of subject ability and the item difficulty. In contrast, the advantage of the Rasch model is that the difficulty of the test item and the ability of the test takers can be independent of the sample; therefore, it is possible to use the linking items to achieve the estimation of the data based on different test papers taken by different groups of test takers on different occasions, and the equating makes comparable on the same scale the ability values (expressed as the test scores) and the difficulty parameters of the test items obtained from different test takers. At the technical level, the Rasch model is the most versatile, powerful, and feasible in objective measurement. The data matrices, regardless of dichotomous or polychotomous in nature, can be processed by Rasch-based software, and the results can be stored in a designated computer file for easy and convenient recall at any time in the future. However, for a long time, these widely used, powerful and feasible methods have been mostly confined within English-speaking countries, while the Chinese researchers of humanities and social sciences and non-English majors have yet to gain a comprehensive understanding of the rationale behind the Rasch model, its application and its efficiency (Andrich, 2004; Bachman and Palme, 1996; Fan, 1998; Engelhard, 2013; Hambleton and Swaminathan, 1985; Kolen and Brenman, 1995, 2004; Linacre, 2002, 2016; Smith, 1990; Shichun Gui, 1986, 1990, 1993; Trevor and Christine, 2015; Trevor et al., 2021; Wright, 1992; Wright and Stone, 1979).

1.9 Rasch Model vs. the One-Parameter Model of IRT[11]

At this stage of the literature review, readers may have found that the Rasch binary-valued scoring model is often referred to as *a one-parameter model* of IRT, and some articles even use the Rasch model and the one-parameter model interchangeably (i.e., Item Response Theory "1-PL", or "1PL"). According to the author, if we look at this from a practical point of view, the explanations of the models are the same, the mathematical formulas used look similar, and the researchers in general are used to the difference in textual expressions. However, should one ask whether there exists any difference between the two, the answer is positive. According to Linacre (2016), Andrich and Ida Marais (2019), Bond et al. (2021), and other Raschers, the difference is conceptual: The Rasch model specifically implements the nature of the additive measure. Data must conform to the Rasch model to support additivity. The IRT model is a "hopeful" description of a set of data. If the data iterations do not converge, another model is tried. According to Stemler and Naples (2021), to put it more bluntly, the major difference between the Rasch model and the one-parameter model is that the former is based on the model per se, while the latter is based on the data. The central problem with the Rasch model is that if the data do not fit the model, then it is the problem of *data*. As for the one-parameter model, if the model and the data are not fit, it is the problem of *model*. In what follows and in view of space, the author will briefly share with domestic scholars the six objective yet significant differences (Linacre, 2016) between the two which are well known but tiresome among foreign counterparts.

1.9.1 The Time When the Models Were Proposed

The Rasch model was first proposed by the Danish scholar Rasch, Georg in 1960. In 1968, Birnbaum, Allan published the article entitled "Certain Latent Trait Models" in Statistical Theories of Mental Test Scores edited by F.M. Lord and M.R. Novick is considered to be the earliest article addressing an IRT model. It is also documented that the founder of IRT was Lord, an American psychometrician, who is credited with proposing the first mathematical model of IRT,[12] in his doctoral dissertation, "Theory of the Test Scores", published in 1951.

1.9.2 Advocates of the Models

The first notable proponent of the Rasch model was Benjamin D. Wright of the University of Chicago, the author of The Best Test Design; the first notable proponent of 1PL of IRT model was Frederic M. Lord, who was working at ETS in the USA. The most active and recognized authority in the Rasch model research field is Professor David Andrich of the University of Western Australia; the most active and recognized authority in IRT research is Professor Ronald Hambleton of the University of Massachusetts, USA.

1.9.3 Conceptual Differences in the Models

As per Linacre (2016), when each test taker in the sample is parameterized for item calibration, it is the Rasch model. When the sample of test takers is parameterized by the mean and standard deviation for item calibration, it is the 1PL of IRT model. One thing that is well known in the Rasch research field, however, is that Georg Rasch himself disagreed that the model he proposed is said to be the 1PL of IRT.

1.9.4 Minimum Reasonable Sample Size

Whether the user is using the Rasch model or the 1PL of IRT model can also be determined in how large a sample size is used. According to Linacre (2016), the minimum reasonable sample size for the Rasch model is 30, whereas Downing (2003) argues that the minimum reasonable sample size for the 1PL of IRT in line with IRT should be at least 200 or more.

1.9.5 Data-Model Matching Problems

For Rasch model users, while conducting parameter estimation and encountering data do not match the model, the users should consider editing the data. In contrast, for IRT users, while calibrating test items and encountering 1PL model does not adequately fit the data, they would consider switching to 2PL or 3PL model.

1.9.6 Formulas of the Models

According to Linacre (2005) and Andrich (2004), the formula for the dichotomous scoring of the Rasch model goes as follows:

$$P_i = \frac{e^{(\theta - b_i)}}{1 + e^{(\theta - b_i)}} \tag{1.5}$$

The 1PL of IRT is formulated as follows:

$$\mathrm{P_i}(\theta) = \frac{\mathrm{e}^{1.7(\theta - \mathrm{b_i})}}{1 + \mathrm{e}^{1.7(\theta - \mathrm{b_i})}} \tag{1.6}$$

Here all the natural logarithmic base is e = 2.71828, the only difference being the additional re-scale factor D = 1.7 in the 1PL of IRT.

As per BILOG User Manual (P.1-2) (Mislevy and Bock, 1990), the 1PL model equation used goes thus:

$$\mathrm{P_{(1)j}}(\theta) = \frac{1}{1 + \mathrm{e}^{-\alpha 1.7(\theta - \mathrm{b_j})}} \tag{1.7}$$

According to Mislevy and Bock, where (e = 2.718) is the base of the natural logarithms, α is a scale constant that determines the ability value θ; b_j is the difficulty parameter associated with item j (i.e., also referred to the item threshold). Items greater than the value of the item b_j are harder items; items with smaller values are easier items.

In practice, these formulas are actually written into the computer program which runs according to the command file. If EXCEL or SPSS is used to do such operations, the users are supposed to edit the relevant mathematical formula and input all the data manually; the process would be extremely cumbersome and error prone. Therefore, the author here just gives the readers a simple tip: if you are not sure whether you are using the Rasch model or the 1PL of IRT, you just check to see which software you are using (see Chapter 3, Section 3.5 for more details).

1.10 Fewer Parameters yet Better Results

Another issue that cannot be ignored in Rasch measurement research and applications is whether to use the Rasch model or IRT. Is it true that more parameters are better? This is a frequently asked question. Due to space, this section borrows the title of Professor Steven Stemler's keynote speech at the

PROMS2021 conference in Nanjing, China[13] to illustrate the advantages of using the Rasch model. The title of Professor Steven's keynote speech is very eye-catching: "Fewer parameters, better results! The real value of the Rasch model over item response theory". This is reflected in the following three broad areas.

1.10.1 Simple Model and Readable Results

The single parameter model used by Professor Steven here refers to the Rasch model, which has only one parameter, thus making the structure of the Rasch model simple to understand and easy to use. For example, in the application of language test, there is only one parameter, that is difficulty; in generating the estimation results, the difficulty value and the ability value (in the same unit, logit) are comparable and interpretable on the same scale, which gives researchers a very intuitive parameter reference.

1.10.2 Objectivity and Reliability

The main goal of the Rasch model is to obtain data that fit the model well, rather than just describing a particular data matrix. This makes it easy for the researcher to edit the data using any Rasch model-based software, especially for binary data types, and even if the data matrix is large enough, the iterations are easy to converge, resulting in objective and reliable estimates.

1.10.3 Widely Used and Easy to Promote

The Rasch model has a wide range of applicability, not only confined within educational measurement, language testing, etc., but also for a wide range of research disciplines such as psychology, medicine, and humanities and social sciences. Once the theoretical ideas and working principles of the Rasch model are understood, it is easy to be accepted by researchers.

Up to now, let's look at this issue from another angle, as per Samuel A. Livingston (2004), who taught a course on test score equating at ETS and used a textbook that explicitly stated on the cover "Without IRT". Samuel's response was simple: While the equating via IRT has become widespread and has important advantages, IRT is conceptually and procedurally complex. According to Samuel, in many similar cases, IRT can be dispensed with because many other methods have simpler procedures, the theoretical basis of which is easier to explain, and the underlying assumptions are closer to reality. This is something that the author knows all too well, just as trying to promote and popularize IRT among the population of College English teaching and testing in China is basically an exercise in futility. There are more examples like this, and because of this, according to Bond (2015, 2016), the most rough estimates, especially for studies published in English, there has been a significant increase in the number of studies applying Rasch's

TABLE 1.3

Number of Publications by Chinese Scholars of Various Disciplines Citing the Rasch Model in Different Disciplines between 1985 and 2016 (by 20 April 2016)

Academic Discipline	Years					
	1985–1989	1990–1994	1995–1999	2000–2004	2005–2009	2010–2016
	Number of Articles					
Book notice	0	1	1	1	0	2
Analytics	1	4	3	3	2	11
Language test	0	0	0	1	11	55
Pedagogical	0	0	0	0	12	61
Psychology	1	1	0	1	15	25
Physical education	0	0	0	0	1	3
Study of medicine	0	1	2	5	5	7
Others	0	0	0	0	2	2
PhD thesis	0	0	0	0	2	13
MA thesis	0	0	0	0	21	44
Subtotal	2	7	6	11	71	223

model since 2000, with more than 500 citations per year, which is five times the number of citations a decade ago. Studies referring to the Rasch model have increased approximately fourfold, doubling between 2001 and 2007 and then doubling again between 2007 and 2013. In order to make some comparisons between these two groups of researchers (i.e., English-speaking and Chinese-speaking researchers), Bond entered the keyword "Rasch Model" into the China National Knowledge Network (CNKN)[14] to show the number of publications by Chinese scholars in different disciplines citing the Rasch model from 1985 to 2016 (as shown in Table 1.3).

From Table 1.3, two points are obvious: (1) the number of citations to the Rasch's model in Chinese scholars' research has increased dramatically since 2005; (2) the number of publications across disciplines in 2016 was more than three times the number in the period 2005–2009. The author argues that such an increase is conducive to popularizing and improving the quality of research and applications of education and language testing in the Pacific-rim countries and regions and even globally. To quote Bond (2016), such a growth rate is faster than that in the Western world. However, it still does not compare favorably with the number of citations to Rasch's model in the Western world, and perhaps an unavoidable problem with this is the lack of sufficient references in Chinese sources.

1.11 Summary

This chapter provides a general introduction to the Rasch model, tracing back to the early "psychometric controversy", reviewing other stories related to the Rasch model, explaining the origins of the Rasch model, introducing the experts and scholars involved in the Rasch research family, and the basic concepts related to the model as well as the examples given by experts and scholars. Pragmatically speaking, all these are by no means the author's original ideas. What the present author actually did is to have rephrased or paraphrased the ideas in the simplest language after referring to a large number of literatures and software manuals or following up with relevant people by WeChat and emails. The purpose is in an attempt to make them easier for novices and neophytes to understand so as to achieve better promotion and dissemination. And in particular, among the profiles elaborated, some practices of which germane to the debate on "standardized testing", the 10-year MET Equating Project implemented in China at an early time first broke in English for publication. The discussions are basically the accumulation of the author's own years of experience in research and application as well as the key member of MET Equating Project, which represents the author's own academic views, and are for the readers' reference only.

Notes

1. See Mike Linacre's (2016) WINSTEPS User Manual for more details; this section is only a brief introduction.
2. See Steven, S.S. (1946). On the theory of scales of measurement. Science, 103, 677–680.
3. The Benjamin Drake Wright Senior Scholar Award is a prestigious accolade given by the Rasch Special Interest Group (SIG) of the AERA. This award is presented biannually to recognize outstanding contributions to the field of Rasch measurement. Named in honor of Benjamin Drake Wright, a prominent figure in the development and application of Rasch measurement theory, the award acknowledges senior scholars who have made significant advancements in research, theory, or practice related to Rasch models. Recipients of this award are typically individuals who have demonstrated a sustained impact on the field through their scholarly work, leadership, and mentorship.
4. MET is abbreviated from Matriculation English Test. In Chinese, it is called GaoKao, the most competitive entrance examination for higher education annually administered by MOE, China to test takers now over 10,000,000. MET is of a large-scale yet high stake examination.
5. Interested readers may refer to Bond (2016, pp. 39–48). Prior to this, research and application of the Rasch model was limited to the Hong Kong Education Examination Department and the Hong Kong Institute of Education in Hong

Kong, China. Mainland China was still largely using traditional analyses based on CTT.

6. No further theoretical account will be covered here. Interested readers may refer to:

   Gui, S. Chuan (1995). Some Reflections on Standardized Tests. Chinese Examination, No. 3.

   Gui, Shichun (1986). Standardized Examinations: Theory, Principles and Methods. Guangdong Higher Education Press.

7. For interested readers, may also see: McNamara, T & U. Knoch (2012). The Rasch Wars: The emergence of Rasch measurement in language testing. Language Testing, 29(4), 555–576. https://doi.org./10.1177/0265532211430367
8. For details, see Peter Tymms (2021).Foreword to the 4th edition of Applying the Rasch Model: Fundamental Measurement in Human Sciences by Trevor G. Bond et al. (2021). Routledge Taylor & Francis Publishing Group.
9. Mathematically, natural logarithm (ln), logarithm with base $e = 2.718281828...$ That is, $\ln (e^x) = x$, where e^x is the exponential function.
10. For details about the similarities and differences inherent in CTT and RMT for dichotomous responses, interested readers may refer to the table addressed by Andrich and Ida Marais (2019, pp. 151–153).
11. Interested readers are referred to http://www.rasch.org/rmt/rmt193h.htm
12. Item Response Theory, usually abbreviated to IRT. In the early days it was also called "Latent Trait Theory", or item characteristic curve theory.
13. Professor Steven Stemler's keynote address was entitled: Better Measurement and Fewer Parameters! The True Value of Rasch over IRT (see Chapter 6, Section 6.5.6 of this book for details).
14. CNKI, http://www.cnki.net/ CNKI is an online data bank supported by the Ministry of Education of China, which provides comprehensive academic resources including journals, doctoral dissertations, master's theses, academic journals, newspapers, and statistical data, etc.

Bibliography

Andrich, D. (2004). Controversy and the Rasch model: A characteristic of incompatible paradigms? Medical Care, 42, 7–16 (Reprinted in E.V. Smith & R.M. Smith, Introduction to Rasch Measurement: Theory, Models and Applications. JAM Press, Minnesota. Ch. 7, pp. 143–166).

Andrich, D. (2019). A Course in Rasch Measurement Theory Measuring in the Education, Social and Health Sciences. Springer Texts in Education. ISBN: 978-981-13-7495-1. ISBN: 978-981-13-7496-8 (eBook). https://doi.org/10.1007/978-981-137496-8

Aryadoust, V., Tan, H.A.H., & Ng, L.Y. (2019). A scientometric review of Rasch measurement: The rise and progress of a specialty. Frontiers in Psychology, 10, 2197.

Atiquil Islam, A.Y.M. (2023). Applying the Rasch Model and Structural Equation Modelling to Higher Education. Boca Raton, London, New York: The CRC Press, The Taylor & Francis Group.

Birnbaum, Allan (1968). Some Latent Trait Models. In F.M. Lord & M.R. Novick (Eds.), Statistical Theories of Mental Test Scores. Reading (pp. 397–497). MA: Addison-Wesley.

Bond, T.G. (2005). Past, Present and Future: An Idiosyncratic View of Rasch Measurement. In Alagumalat et al. (Eds.), Applied Rasch Measurement: A Book of Examplars (pp. 329–341). Netherlands: Springer.

Downing, S.M. (2003). Item response theory: Applications of modern test theory for assessments in medical education. Medical Education, 37, 739–745.

Engelhard, G. (2013). Invariant Measurement Using Rasch Models in the Social, Behavioral, and Health Science. New York and London: Taylor & Francis Group.

Engelhard, G., & Wang, J. (2024). Invariant Measurement Using Rasch Models in the Social, Behavioral, and Health Sciences (2nd Ed). New York and London: Taylor & Francis Group.

Fan, X. (1998). Item response theory and classical test theory: An empirical comparison of their item/person statistics. Educational and Psychological Measurement, 58(6), 895–908. https://doi.org/10.1177/0013164498058006001

Hambleton, R., & Swaminathan, H. (1985). Item Response Theory: Principles and Applications. Kluwer: Academic Publisher.

Keeves, J.P. (1997). International practice in Rasch measurement, with particular reference to longitudinal research studies. Invited paper presented at the annual meeting of the Rasch Measurement Special Interest Group. Chicago, IL: American Educational Research Association.

Linacre, J.M. (2002). What do infit and outfit, mean-square and standardized mean? Rasch Measurement Transactions, 16(3), 878.

Linacre, J.M. (2005). Rasch dichotomous model vs. one-parameter logistic model. Rasch Measurement Transactions, 19(3), 1032.

Linacre, J.M. (2016). A User's Guide to WINSTEPS Rasch Model Computer Programs: Programme Manual 3.92.0[EB/OL]. ISBN 0-941938-03-4.3(4); 78. https://www.winsteps.com/winman/copyright.htm

McNamara, T., & Knoch, U. (2012). The Rasch wars: The emergence of Rasch measurement in language testing. Language Testing, 29(4), 555–576. https://doi.org./10.1177/0265532211430367

Mislevy, R.J., & Bock, R.D. (1990). PC-BILOG: Item Analysis and Test Scoring with Binary Logistic Models. [Computer Software] (2nd Ed.). Scientific Software, Incorporated P.O. Box 536. Mooresville, IN 46158 USA.

Mok, Magdalena M.C., & Quan, Zhang. (2018). Chinese editors of 《罗氏测量：应用与导读》. Original English version edited by Everett V. Smith, Jr. Richard M. Smith. JAM Press, U.S.A.

Rhodes, Novick. (1992). The Statistical Theory of Psychological Test Scores, Fujian Educational Publishing House. Original work by Lord, Frederic M., & Novick, Melvin R. (1992). Statistical Theories of Mental Test Scores. ISBN 9787533407483

Rasch, G. (1960/1980). Probabilistic Models for Some Intelligence and Attainment Tests. Expanded edition (1980) with foreword and afterword by B. D. Wright (Ed.). Chicago: The University of Chicago Press. Reprinted (1993) Chicago: MESA Press.

Steiner, Steven. (2021). "Better Measurement and Fewer Parameters! The True Value of Rasch over IRT", The keynote speech at PROMS2021, Nanjing, China.

Stemler, S.E., & Naples, A. (2021). Rasch measurement v. item response theory: Knowing when to cross the line. Practical Assessment, Research and Evaluation, 26(1), 11.

Sullivan, J.I. (1988). Editor's Introduction. In D. Andrich (Ed.). Rasch Models for Measurement (pp. 7–8). Newbury Park, CA: Sage. http://doi.org/10.4135/9781412985598

Thurstone, L.L. (1928). Attitudes can be measured. The American Journal of Sociology, 33(4), 529–554.

Thurstone, L.L. (1959). The Measurement of Values. Chicago: University of Chicago Press.

Trevor, G.B., & Christine, M. Fox. (2015). Applying the Rasch Model: Fundamental Measurement in the Human Science (3rd Ed.). Routledge Taylor & Francis Group.

Trevor, G.B., Yan, Z., & Moritz, H. (2021). Applying the Rasch Model: Fundamental Measurement in the Human Science (4th Ed.). New York and London: Routledge Taylor & Francis Group.

Trevor, G.B., & Yan, Zi. (2016 [2017]). Exporting to China: The Future of a Genuine Collaboration with the West. In Q. Zhang (Ed.), Pacific-Rim Objective Measurement Symposium PROMS 2016 Conference Proceedings (pp. 39–48). ISBN: 978-981-10-8137-8; ISBN: 978-981-10-8138-5 (eBook). http://doi.org/10.1007/978-981-10-8138-5

Wright, B. D. (1992). IRT in the 1990s: Which models work best? Rasch Measurement Transactions, 6(1), 196–200.

Zhang, Q. (2019). Rasch Model: Research and Practice in China. In Myint Swe Khine (Ed.), International Trends in Educational Assessment Emerging Issues and Practices. Brill. ISBN 978-3-662-47489-1. Available at http://www.catalog.loc.gov

Zhang, Q. (Ed.). (2016). Pacific-Rim Objective Measurements Symposium PROMS 2015 Conference Proceedings. ISBN: 978-981-10-1686-8; ISBN: 978-981-10-1687-5 (eBook). http://doi.org/10.1007/978-981-10-1687-5

Zhang, Q. (Ed.). (2017). Pacific-Rim Objective Measurement Symposium PROMS 2016 Conference Proceedings. ISBN: 978-981-10-8137-8; ISBN: 978-981-10-8138-5 (eBook). http://doi.org/10.1007/978-981-10-8138-5

Zhang, Q., & Yang, H. (Eds.). (2013). Pacific Rim Objective Measurement Symposium PROMS 2012 Conference Proceedings. ISBN: 978-3-642-37591-0; ISBN: 978-3-642-37592-7 (eBook). http://doi.org/10.1007/978-3-642-37592-7

Zhang, Q., & Yang, H. (Eds.). (2015). Pacific-Rim Objective Measurement Symposium PROMS 2014 Conference Proceedings. ISBN: 978-3-662-47489-1; ISBN: 978-3-662-47490-7 (eBook). http://doi.org/10.1007/978-3-662-47490-7

桂诗春、李崴、张权. (1993). 项目反应理论在高考成绩等值上的应用.第4届全国教育考试科研讨论会. ISBN 7-80101-089-2/g.64

桂诗春. (1990). 题库建设讲话 (II). [J]. 《现代外国语》 (4), 66-72.1990 (1):66-72,59.

2

Basic Concepts in Classical Testing Theory

Today, classical testing theory (CTT) remains one of the foundational theories in the assessment of psychological and educational tests, providing essential tools for understanding and improving the reliability and validity of tests. Its long history tracing back to the early 20th century witnessed the idea of factor analysis and reliability proposed by Charles Spearman, the principles of test construction and the need for objective measurement by Edward Thorndike, the concept of multiple-factor analysis by Louis Leon Thurstone, Kuder and Richardson's Formula 20 (KR-20), a measure of internal consistency reliability for dichotomous items, and Lee J. Cronbach's alpha, a widely used measure of internal consistency reliability, all of which are the essential tools used in CTT today.

After WWII, CTT became a dominant framework in psychometrics, guiding the formation of standardized tests in educational settings. The key concepts of CTT are typically addressed in five aspects: true score theory, observed score, reliability, error of measurement, and validity.

According to CTT, each test taker's observed score on a test falls into two parts: the **True Score** and the **Error Score**. The true score stands for the actual level of the latent trait being measured, while the error score is referred to as the random measurement errors that can affect the observed score. The **Observed score** is what is usually called "raw score" that is the score actually obtained by a test takers on a test and that is also the sum of the true score plus the error score (see Section 2.1.9). In CTT, **reliability** is meant the consistency of test scores based on repeated administrations of equivalent test items, using Cronbach's alpha to indicate the proportion of variance in observed scores due to true score variance. On the other hand, random **measurement error** is assumed to have a mean of zero across a large number of test administrations. Finally, we have **validity**, which is referred to as the extent to which a test measures what it should be to measure. In CTT, validity is another crucial aspect regarding test evaluation. Despite its widespread use and being influential, limitations remain (see Section 1.8). These limitations led to the development of more sophisticated models, such as Rasch models and IRT models. The latter addresses the limitations inherent in CTT by focusing on the relationship between test items and latent traits (see Chapter 3).

To make the content of this book easy to understand, the author will try to use simple formulas or express in words rather than using complex calculation formulas and processes to focus on introducing basic concepts in CTT mostly germane to Rasch-GZ. These basic concepts will be addressed

DOI: 10.1201/9781003665595-2

in two parts: item analysis (IA) used in the RASCH-GZ system and topics about standardized examinations. All these concepts are the foundations for understanding language testing in general and IA in specific. The basic concepts addressed in this chapter can be basically found in most books on psychometrics and educational statistics and will provide useful references for other chapters. For a more detailed discussion on CTT, readers can refer to the references listed at the end of this chapter.

2.1 Item Analysis (IA) Defined

IA is an important aspect of test development process. It offers criteria for good test items, or rather test items should meet specific technical assumptions based on IA. In CTT, IA (Gui Shichun, 1986; Lynch, 2003) refers to the internal evaluation of test items, which involves analyzing a series of parameters of each test item in a test paper one by one (as will be discussed from Sections 2.1.1–2.1.14). They include correct answer rate, P-value, difficulty, score distribution, normal and abnormal distribution, mean and standard deviation (SD), correlation, skewness, kurtosis, range, reliability, α value, Δ value, and discrimination (including discrimination analysis for distractors in multiple choice questions). On the basis of these analyses, one can form an opinion on how important IA is in the practice of CTT.

2.1.1 Importance of Item Analysis

In the research and practice of language testing, a universal phenomenon is that not enough attention has been attached to IA. For a long time, no pretest is conducted in schools. The common practice is that once an exam is over and the scores are used, the test paper will be put aside and ignored. In fact, IA has significant value in at least the following two aspects. In the first aspect, IA analysis shows how test takers responded to each test item in an entire test paper. For example, if some test taker(s) declare that a certain test paper is difficult or easy, we needn't to agree or disagree instantly. Instead, we should check to see if the actual results are based on a big size of sample population or not. That is to say, we may refer to data and will not make any decision abruptly based on the subjective voices from certain test takers or schools. This is extremely important for the practice of large-scale examinations with high stakes. Next, IA calibrates each test item and generates both the difficulty and discrimination of each test item. This is crucial for testing personnel and examination management departments. That means we not only know the difficulty level of all the test items in a timely manner but also know the quality of all the test items, i.e., how well each test item was actually designed, including key answers and distractors in a multiple-choice questions (MCQs). They serve as valuable feedback for further improvement of the test paper.

These are the core methods in CTT, as any individual analysis often deviates from the actual results. In the era of big data, everything should be based on objective quantities. In doing so, we need to know more about CTT, and let us start from pre-test and item moderating first.

2.1.2 Pre-test and Test Item Moderating

At the beginning of this chapter, the author reminds readers that the concept of *"pretest"* in CTT is different from that used in general statistics. The latter refers to an experiment conducted before the start of research or pre-test vs. post-test to check the reliability between two groups of data. In CTT, a pre-test is actually referred to as the practice of administering a test paper to a sample of target population to collect data for conducting item analyses. With the item analyses based on the pre-test data, the test designers would achieve purposes of twofold.

One purpose is checking whether the overall difficulty level, the structure and the proportion of test items used in a test paper, and even the time for test performance are appropriate for future candidates. The other purpose is to moderate test items. In CTT, item moderating refers to the professional practice to check the quality of the test items based on the parameters provided by both the item analyses and a detailed report. Specifically, this includes improving, replacing, or discarding any test item that does not meet the requirements based on the IA in terms of difficulty and discrimination (including both key and all the options). These are the two important measures taken by test designers to ensure internal consistency, to reduce external interference, and to avoid any biased items, thus making the overall item difficulty of the test paper meet or approximate to the actual ability level of the targeted test takers. The role of pre-test and moderation in CTT is crucial in specific implementation of exams, especially in large-scale and high-risk exams. They not only provide a reliable basis for exam designers and testing professionals to evaluate the test paper but also serve as an important foundation for drawing scientific conclusions.

For example, in the case of implementing a high school entrance examination, it's not merely organizing a group of testing experts to produce a test paper with answer keys based on the exam syllabus and then administering the test right way. In CTT, the common practice is to first design a set of test papers according to the exam syllabus and administer this pre-test to a sample group of the targeted schools. The pre-test sampling should be administered in the same setting as the real entrance examination under the premise of avoiding the "Hawthorne Effect". After the pre-test, the data are collected to conduct IA and testing scoring. With item analyses as discussed above, the test designers will do item moderating. The examination to be administered to the whole population should be well moderated. This is what we call "pre-test" in CTT.

2.1.3 Correct Answer Rate, P-value, and Difficulty

In CTT, the correct answer rate actually refers to the P-value, which is the ratio of test takers who answered a question correctly to all the test takers in a test (R/T). From another perspective, it also reflects the difficulty level of the test item. The idea is usually expressed in the following simple form:

$$P = \frac{\text{Number of test takers choosing correct answers}}{\text{Total number of test takers in one exam}} \tag{2.1}$$

According to Brian (2003), this is just a different interpretation of the information. The larger the P-value, the easier the test item for the test takers. Therefore, some scholars (Andrich, 2019) also refer to it as the "Facility Index" (i.e., the ability for answering test items correctly). In dichotomously scored items, the index is simply the proportion of persons who answered the item correctly and usually denoted by P.

In CTT, P-value is also interpreted as difficulty. A test item with a P-value of 0.5, meaning that 50% of test takers get the correct answer, is generally considered "ideal test item". Test items with extremely high (or low) P-values will be considered as requiring moderation.

However, in dealing with the "ideal difficulty" for a test paper, one important point should be taken into consideration. Gui, Li, and Li (as cited in Xiao, 1988), if the test paper consists of MCQs with four options, the guessing factor (1/4) should be considered as well. Therefore, in CTT, the difficulty level is usually increased to 0.625 according to the formula that excludes guessing factors:

$$\text{The revised correct answer rate} = \frac{\text{The actual wrong answer rate}}{\text{the number of options} - 1} \tag{2.2}$$

Therefore, we have the revised correct answer rate is $0.625 - 0.375/3 = 0.5$.

In the Rasch GZ system, we design, in large-scale examinations with high stakes, test items that only 10% of test takers answer correctly as Very Difficult items, i.e., ($VD < 0.1$), that 10–30% of test takers answer correctly as Difficult items, i.e., ($D = 0.1 \approx 0.3$), that 30–70% of test takers answer correctly as Intermediate items, i.e., ($I = 0.3 \approx 0.7$), that 70–90% of test takers answer correctly as Easy items, i.e., ($E = 0.7 \approx 0.9$), and that more than 90% of test takers answer correctly as Very Easy items, i.e., ($VE > 0.9$).

2.1.4 Score Distribution

In CTT, score distribution mainly refers to the number (or percentage) of test takers at each score level. According to Gui Shichun (1986), the first test to be conducted on an exam is to see if the distribution of its scores is normal. The individual differences among test takers participating in large-scale exams are relatively large, and it is crucial for the exam to accurately reflect these differences and test takers' different levels as different scores. According to the principles of statistics, so long as the sample taken is large enough, its distribution will exhibit normality. In large-scale exams, the distribution of candidates' levels (abilities) should be normal, and so should be the distribution of exam scores. In case a serious deviation from the normal score occurs, it is often due to the fact that the level of population is normal, but the exam score does not accurately reflect the level of the population. For example, regardless of the type of exam, there will exist test takers with upper, middle, and lower levels, respectively. However, due to the easy presentation of the test paper or the lenient grading criteria, test takers with intermediate ($I = 0.3 \approx 0.7$) or higher levels ($VD < 0.1$) and ($D = 0.1 \approx 0.3$) will be given high scores, making it difficult to distinguish between upper and intermediate level test takers in the exam, resulting in a negative bias (see Section 2.5). On the contrary, if the test paper is difficult or the grading criteria are strict, test takers below the average level are given low scores; thus, it is difficult to distinguish between test takers at the average and lower levels in the exam, resulting in a positive bias. By examining the distribution of scores, we can often understand the quality of test items used in exams. This is a fundamental concept in language testing.

Another thing worth mentioning is that when we talk about score distribution, we may have two types of exams in mind: norm-referenced and criterion-referenced tests. These two types of exams are associated with score distribution.

2.1.4.1 Norm-referenced Test (NRT)

In CTT, exams can be divided into norm-referenced and criterion-referenced according to the way the scores are demonstrated. A test of norm-referenced refers to a certain norm that is taken as reference to reflect the scores of test takers. Here, the **norm** is interpreted as the scores achieved by a group of test takers of homogeneous in nature (each of whom is one of them) in an exam statistically expressed in mean (M) and standard deviation (SD). So, the NRT is actually reflecting a test taker's score with the scores achieved by other test takers for reference. This is particularly necessary in large-scale exams aimed at selecting the best students because selection here means to make decision based on comparison. However, it is inappropriate to make a decision based solely on a candidate's score, regardless of others' cases. In China, school teachers may evaluate students from two perspectives: one is the raw scores they receive on the test paper, and the other is the order in which the scores are ranked according to the actual level in the class. For example, if a student gets 85 points and ranks the 3rd in the class, we know that these 85 points belong to a higher level in the class, and only two other scores are higher than this one. However, in the case of large-scale exams with thousands of candidates participating, such a scoring practice is infeasible to

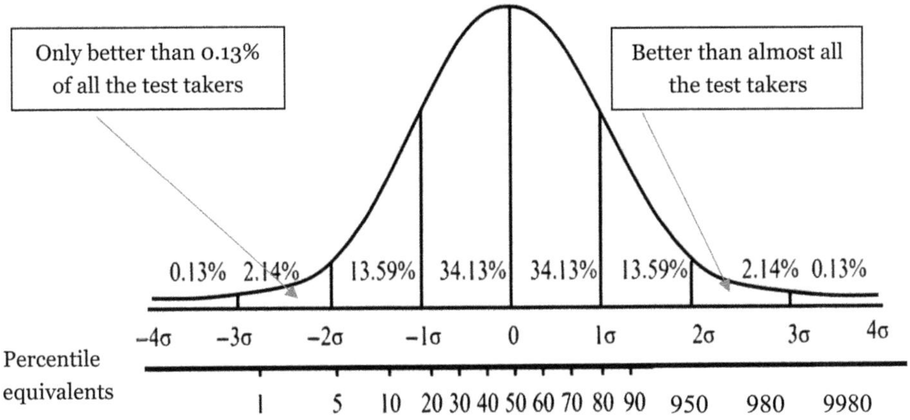

FIGURE 2.1
The ideas of norm-referenced evaluation

achieve, and it is also difficult to figure out the meaning of scores. Let's take the above example again. If the candidate still scores 85 points, but his/her order is ranked at 983, is this score considered high or low? Our answer is that it depends on the total number of test takers. Suppose we have a total of 1,000 test takers, this score should be evaluated at the bottom line; if the total number of test takers is NOT 1,000, but 10,000, then this score is relatively much higher. Therefore, in CTT, instead of ranking an order, it would be better to give a percentage: if the total number of candidates is 1,000, then the position of this candidate is at 93.8% or so, which indicates lower than 93.8%, or only higher than 6.2% of test takers; if the number of candidates is 10,000, the position of the student is at 9.38%, which is higher than 90.62% of candidates. The method of reflecting a candidate's score in the norm-referenced exam is precisely to evaluate a test taker's score by referring to the situation of other candidates, which is particularly beneficial for selecting students. The ideas are illustrated in Figure 2.1.

2.1.4.2 Criterion-referenced Test (CRT)

If a norm-referenced exam must refer to a norm (i.e., the average score based on all test takers) when grading, then a criterion-referenced exam is just the opposite (Popham, 1997, 1994, 1978). The practice of a test of criterion-referenced abides by a predetermined scale or goal when grading. That's why testing professionals also call it a target-referenced exam. If the scores given in NRT are relative, the scores given in CRT are fixed, meaning CRT does not take into account what scores are achieved by other test takers. In other words, CRT is not about picking the tall from the short, but about setting a certain scale or height, and only by reaching this height can one pass. Therefore, for CRT, if all candidates reach the predetermined level, they should all pass; if no candidate reaches the level, none should pass. In this sense, the tests of high jump and long jump in school physical education courses should be CRT. In society, some tests such as drivers obtaining driver's licenses, lawyers or accountants obtaining business certificates, all of which must use CRT. In CTT, unless there is a special need, in general, achievement tests (ATs) refer to CRT, while proficiency tests (PTs) refer to NRT.

2.1.4.3 Achievement Test (AT)

From the perspective of purpose, language tests can be divided into AT and PT (Gui, 1986). In CTT, an AT is used to measure what students have actually learnt from a textbook based on teaching syllabus over a period of time. AT checks the learning progress and must echo past teaching. Test items thus designed in AT cannot be separated from what was taught as indicated by solid arrows in Figure 2.2. The score obtained from AT is used to indicate the degree to which students have mastered the teaching content. If a percentage scoring is used, getting 60 points often means mastering 60% of the teaching content. The score of AT is often determined by the two aspects: (1) the amount or depth of teaching content. If the teaching content is limited or shallow, students are more likely to score high, or vice versa. (2) The level of mastery of teaching contents. In CTT, the scores of AT between schools are not comparable due to different textbooks, progress, and testing methods. Evaluating students' level based on AT scores is often inaccurate due to the lack of a unified standard. From the perspective of students' scores, the pass rate of AT should be relatively high. A low pass rate can bring many problems to schools and society. In a word, AT is a type of local examination based on textbooks, often administered by teachers who taught the class themselves. Mid-term and final exams as well as graduation exams on campus all belong to AT.

2.1.4.4 Proficiency Test (PT)

The PT is used to measure whether a test taker has reached level X, in order to determine whether they are suitable for a future task Y. As can be seen in Figure 2.2, though X is related to past teaching, it does not directly correspond

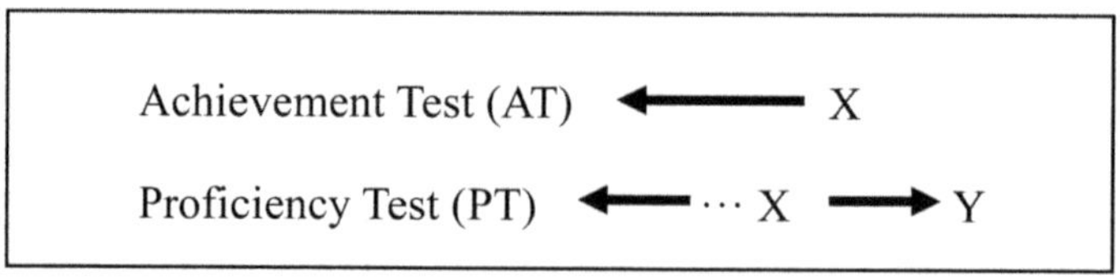

FIGURE 2.2
Schematic diagram of the purposes of AT and PT

to it; therefore, it is represented by a dotted line. According to Gui (1986), the PT should take into account what the students have learned (or should have learned). If not considered, the purpose of the test will not be achieved. However, the PT test items cannot be based on any textbook because the PT attempts to use a unified standard to test schools (students) using different textbooks. Using a particular textbook as a basis will benefit only schools that use that textbook. The PT test item writing not only takes into account the students' learning over the past but also the future task Y. For example, English PTs, such as WSK[1] in China, TOEFL in the United States and Cambridge English Proficiency Test in the UK aim to measure whether a candidate's English proficiency can meet the needs of learning and living in English speaking countries, or Matriculation English Test (MET) in China will not only consider students' learning in the middle schools but also for higher education after entering universities. The test item design and production of PT must be based on the practical needs. The PT scores can be compared because the test will not be biased toward any particular school, or university. Generally speaking, the PT score does not indicate whether it is a pass or a fail. It only gives the test taker a score with reference to other scores of the same test. Now we can see, if a test is of selective, it must be of proficiency, as they all require completing certain tasks and can only select the best candidates from the same population. Therefore, the PT requires a higher ability to distinguish among candidates of different levels, such as upper, middle, and lower. The MET in China should be a type of PT.

2.1.5 Normal Distribution and Abnormal Distribution

In language tests, when it comes to test takers' scores, abilities, or item difficulty levels, the terms "normal distribution" or "abnormal distribution" are often encountered.

According to Zhang Houcan (1986), the normal distribution is a type of continuous, random variable, probability distribution. Simply put, it is a normal (curve) distribution, or rather a type of frequency distribution. Normal distribution is a theoretical distribution that plays an important role in language testing theory and practice.

From a statistical perspective, a normal distribution has the following three characteristics:

(1) The most intuitive manifestation of a normal distribution is its symmetrical bell-shaped curve (see Figure 2.1), characterized by the mean as the axis of symmetry, with peak values corresponding to the mean. As the distance from the mean increases, the curves at both ends gradually decrease, approaching (∞) zero infinitely, but never zero.

(2) A normal distribution is a unimodal distribution, where the density function reaches its maximum value at the mean. We can also

understand it as follows: the data is most concentrated near the mean where the probability is higher. The farther away from the mean, the lower the probability of data appearing. The idea represented by percentage is best illustrated in Figure 2.3.

(3) The normal distribution is determined by two parameters: mean (μ) and standard deviation (σ). The mean (M) determines the center position of the distribution, and the standard deviation (SD) determines the degree of dispersion of the distribution, which can be expressed as $X \sim N(\mu, \sigma^2)$, where μ is the M and σ^2 is the variance. When $\mu = 0$, when $\sigma = 1$, it is called a standard normal distribution, usually represented as N (0, 1). As shown in Figure 2.1, in the normal distribution, the mean, median, and mode are the same, that is, mean = median = mode = 0. At this point, the concept density value is the highest (y = 0.3989) as the red arrow indicates in Figure 2.1 (Zhang Houcan, 1986).

In addition, Figure 2.1 also indicates the SD and the corresponding area under the normal curve. According to Li Shaoshan (1999), if we want to know the area corresponding to an observation value in a normal distribution, or if we know the area, we don't need to calculate the corresponding observation value anymore. We just need to convert the observation value into a standard score and look up in the normal distribution table to get the answer.

Statisticians generally believe that the concept of normal distribution was first introduced by French mathematician A. D. Moivre and named after scholar C. F. Gauss's in-depth study of its application in error analysis in the 19th century. Therefore, in statistics, normal distribution is also known as

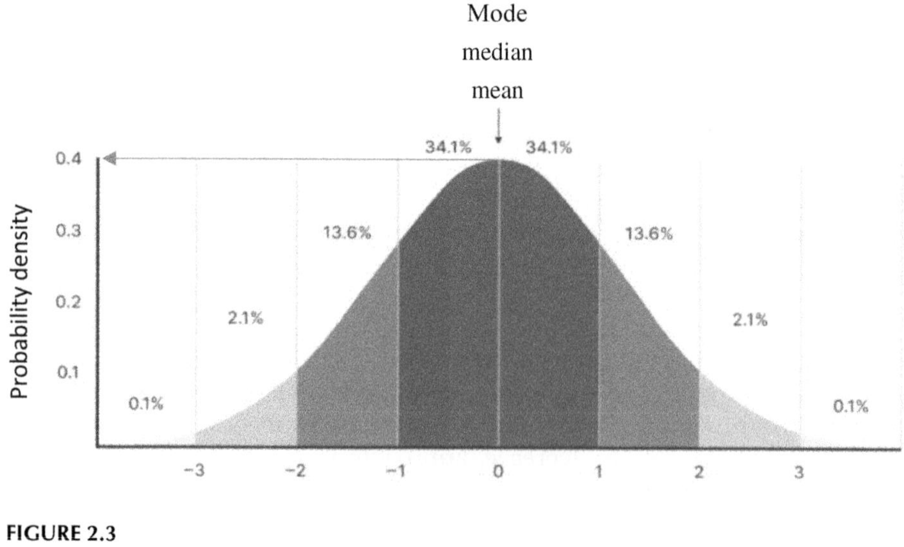

FIGURE 2.3
Standard normal distribution

Gaussian distribution. As for abnormal distribution, it is generally deemed that when the data does not conform to the normal distribution, such that skewness or kurtosis issues occur (see Section 2.6), the traditional normal distribution model is not applicable.

2.1.6 Mean, Variance, and Standard Deviation (SD)

(1) In language testing, when we talk about Mean, we are basically referring to the average score, usually represented by (M). In CTT, M is calculated by adding up the scores of all test takers and dividing by the number of test takers. The basic formula is as follows:

$$M = \frac{\sum \text{Sum of scores of all test takers}}{\text{Total number of test takers}} \tag{2.3}$$

In the above equation, the Greek letter $\sum$ (sigma) represents the sum of all scores.

(2) Variance, also known as mean square. For better understanding, this section uses the symbol S^2 to represent variance. By variance is referred to as the mean of the summed differences between each data point and the mean of the group multiplied by the square (also understood as the distance from the mean), i.e., the M after being squared. In CTT, variance is an important indicator of the dispersion of test data, and its basic formula goes as follows:

$$S^2 = \frac{\sum (X - X)^2}{N} \tag{2.4}$$

Variance helps us understand the distribution of exam data. The larger the variance, the more dispersed the data distribution is; the smaller the variance, the more concentrated the data (refer to Section 2.5, skewness and peak values).

(3) Standard deviation, denoted as SD for ease of understanding. The basic formula for SD is as follows:

$$SD = \sqrt{\frac{\sum (X - X)^2}{N}} \tag{2.5}$$

From the above formula, we can see that the SD is actually the square root of the variance, which can also be expressed as

$$SD = \sqrt{S^2} \tag{2.6}$$

In CTT, an important application of SD is to standardize the raw scores for the purpose of comparing test scores. This is extremely useful for both language testing and teaching. Here, the author further uses the simplest language to expand and explain the meaning of $\sum(X-\bar{X})^2$ in the root of the above formula. This formula actually tells us the following specific steps:

(1) The Greek letter $\sum$ above represents the sum of all fractions;
(2) X represents a particular raw score;
(3) X bar represents the M based on all original scores;
(4) Calculate the difference between all raw scores and the M of the original scores, which is the "deviation from M" mentioned in (Section 2.3.2);
(5) Work out the square of all deviations from the mean;
(6) Add up all squared deviations, that is, to find the sum of all squared deviations;
(7) Divide the sum of squared deviations by N, i.e., the total number of test takers;
(8) Obtain the variance, and the square root of the variance, i.e., the standard deviation.

Hopefully, this will clarify the relationship between what is variance and what is SD for most Chinese teachers of English and humanities in domestic universities. To help further understand, let's use a small group of numbers to illustrate the ideas above.

Suppose we had a classroom quiz consisting of 10 test items to 15 students and obtained the following 15 scores as indicated in Table 2.1.

2.1.7 Correlation

Correlation is a very important concept in statistics, usually represented by r, used to measure the strength and direction of the relationship between the scores of the same test taker in two tests.

There often exists a certain degree of connection or correlation between variables, for one example, the value of variable X may increase with the increase of variable Y, or decrease with the increase of variable Y. For another example, we have found in teaching or testing that students who perform well in dictation often have a higher overall level of English proficiency. Such regular or systematic relationship between variables is called *Correlation*. Correlation essentially refers to the covariation between variables, where one variable changes with the variation of another variable. Since there is a regular relationship between related variables, having the value of one variable can to some extent predict the value of another variable. The accuracy of prediction obviously depends on the strength of the correlation between

TABLE 2.1

The Procedures to Obtain Variance and SD

No.	Score	M	Score-M	(Score-M)^2
1	8	5.133	2.87	8.28
2	9		3.87	14.95
3	4		−1.13	1.28
4	5		−0.13	0.018
5	6		0.867	0.75
6	7		1.867	3.48
7	9		3.867	14.95
8	2		−3.13	9.82
9	3		−2.13	4.55
10	2		−3.13	9.82
11	4		−1.13	1.28
12	1		−4.13	17.08
13	9		3.867	14.95
14	7		1.867	3.48
15	1		−4.13	17.08
Total	77			121.733329
Variance				8.11
Check square root table to obtain				$\sqrt{8.11}$
SD				2.85

variables. If two variables are completely correlated (which is a rare situation), the accuracy of prediction can reach 100%.

2.1.8 Skewness and Kurtosis

Skewness and kurtosis are two other important indicators reflecting the symmetry of the data distribution, with a Skewness value of 0 indicating a normal distribution of data, item difficulty, score, etc. Skewness value > 0 indicates a positive skewness, indicating that the test is difficult in the RASCH-GZ system; a skewness value < 0 indicates a negative skewness, indicating that the test is relatively easy.

Kurtosis, meaning peak value, reflects the concentration or flatness of the data distribution. Kurtosis = 0 indicates a normal distribution; Kurtosis > 0 indicates a concentrated and sharp distribution; Kurtosis < 0 indicates a flat and dispersed distribution. The larger the kurtosis, the more data is generally concentrated around the center point of the coordinate image. The smaller the kurtosis, the flatter the data distribution, and it spreads out toward both

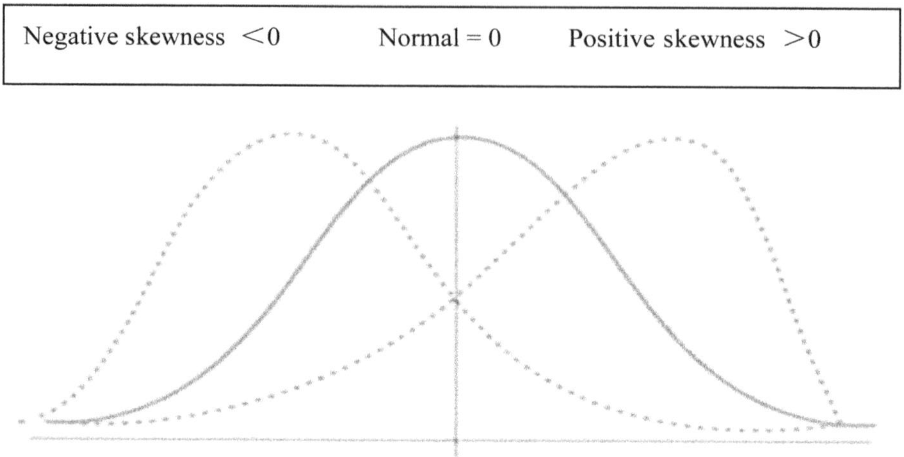

FIGURE 2.4
The curves of positive, negative skewness, and normal distribution

sides (flat) at the center point of the coordinate axis. For general users, with the concepts of skewness and kurtosis, it is possible to infer the distribution of data without seeing the data distribution map. Figure 2.4 best illustrates the ideas.

2.1.9 Range

In CTT, the range, also known as the difference between two extremes, is an important concept of the disparity, generally represented by the letter R. The range refers to the difference obtained by subtracting the lowest score from the highest score in a set of test data, and is used to measure the degree of dispersion of test scores. The calculation method for full range is simple and intuitive, that is:

$$R = \text{highest score} - \text{lowest score} \tag{2.7}$$

According to the kurtosis (2.6) section, the full range can only be used as a rough indicator to infer the data distribution.

2.1.10 Reliability

In CTT, reliability is also known as consistency. According to Professor Gui Shichun (1986), when we say that a measurement is completely reliable, we are actually referring to such a fact that the measurement excludes all errors and is absolutely accurate. That is to say, as long as the measured object itself remains unchanged, using the same "ruler" to measure will always yield the same result. However, the object of educational measurement is people, not objects, and human factors are unstable, making it almost impossible to completely eliminate errors. So, the scores (X) obtained from test takers in exams often contain some errors (e); only by excluding errors is its true score (t), and the relationship among these three can be expressed as the formula as follows:

$$X = t + e, \text{ or } t = X - e \tag{2.8}$$

Generally speaking, true scores are assumed numerical values that are difficult to measure directly. If we want to understand someone's certain ability, we can repeatedly measure them an infinite number of times, and the score measured each time will have an error, but the average of the scores measured all times is equivalent to their true score. In the real testing context, we cannot constantly measure repeatedly, so the true score is but a hypothetical score; however, it is very important for estimating reliability. In any testing practice, errors are inevitable, and researchers need to try to minimize them as much as possible to approach the true score. In CTT, this is the essential meaning of "True Score Theory" and also the classic reliability theory. It began to rise in the late 19th century, formed a relatively complete system in the 1930s, gradually matured, and achieved the transformation to modern measurement theory (Gui, 1986).

There are many formulas for calculating reliability. In the IA used by RASCH-GZ system project analysis, the most accurate reliability coefficient formula (Kuder-Richardson 20) is selected as follows:

$$r_{tt} = \left(\frac{n}{n-1}\right)\left(\frac{\delta_t^2 - \sum pq}{\delta_t^2}\right) \tag{2.9}$$

In the above equation:

n = number of test items
p = Correct answer rate
δ_t^2 = variance of the entire test paper q = (1-p)

The above formula is only suitable for calculating the reliability of objective test items mainly because objective test items still play a dominative role in large-scale exams with high stakes. If subjective test items (such as essay writing, short answer questions, etc.) need to be considered for reliability calculation, Cronbach's alpha coefficient formula is more commonly used.

2.1.11 α Value

In the Rasch-GZ system, the α value is the Cronbach's alpha coefficient value. This is an important indicator for measuring the internal consistency and reliability of a test paper. The range of α values is generally between

0 and 1, with higher values indicating better internal consistency reliability. However, the specific value of α to be reached may vary depending on different research fields, research purposes, and requirements. When using the Rasch GZ testing system, we set the α value to 0.8 or above for effectiveness. There are two points to note here: (1) The value of α is affected by the sample size, and the larger the sample size, the higher the value of α is usually required; (2) The alpha value is the minimum limit of reliability, and the alpha values calculated using the alpha value reliability formula are relatively low (Gui, 1986).

2.1.12 Δ Value

In mathematical statistics, the value of Δ (pronounced as delta) is another very important concept, usually represented by the Greek letters Δ (uppercase) or δ (lowercase) to indicate the difference or change between two variables. Δ value has wide applications in many practical problems, and its specific significance may vary depending on the research field. The Δ value selected for RASCH-GZ was originally a scale recommended by Educational Testing Service (ETS) in the United States:

$$\Delta = 13 + 4Z \tag{2.10}$$

where the constant 13 (equivalent to 0 of Z value, meaning half of the test takers got correct answers) and the constant 4 (equivalent to 1 of Z value, meaning that according to a normal distribution, exactly 84.13% of the test takers answered correctly) are arbitrarily determined. The purpose is to remove negative values and ensure to obtain integers as much as possible. A test item that almost 100% (such as 99.87%) of test takers answered correctly falls at the −3δ position of the normal distribution, D = 13 + (4) (−3) = 1. It should be the easiest item. At the other end of the distribution, a test item answered correctly by fewer than 1% (such as 0.13%) falls at the +3δ position of the normal distribution, D = 13 + (4) × (3) = 25. It should be the most difficult item. Therefore, the difficulty represented by the ETS scale in the United States ranges from 1 to 25. The higher the value, the greater the difficulty. For example: P = 0.41, converted to Z value=0.23, then converted to D value, that is, D = 13 + 4 × 0.23 = 13.92. The P-value of D-value being 13 is 0.5. Now the D-value is slightly greater than 13, indicating that the difficulty of this test item is slightly greater than 0.5 (Gui, 1986).

2.1.13 Concept of Discrimination

In CTT, the concept of discriminative index plays an important role in detecting the difficulty of correct answers and the distractor interference of MCQs. However, in the author's opinion, the idea of "discrimination index" remains somewhat foreign to most Chinese teachers of English. According to Popham

(1990, p. 277), "Item discrimination" describes the relationship between a test taker's answer to an item and their answers in the overall test. That is to say, discrimination answers such a question: If a test taker scores high in the overall test, did they also correctly answer this (relatively difficult) question? Andrich (2019) expressed the similar idea. The term discrimination is used to indicate a statistical index that describes the degree to which an item is consistent with the other items in helping distinguish the proficiencies of the persons. The index is also used to summarize such an expectation: If an item assesses the same variable as the majority of the items of an instrument (a test), then those who obtain a high score on the item, should also tend to obtain a high score on the test. Likewise, those who obtain a low score on the item should tend to obtain a low total score on the test.

When the test items are of MCQ type, discrimination can also be applied to analyze the design level of each distractor. This technique involves calculating the number of test takers choosing each option in the multiple-choice answer format. In other words, it is to what extent the "distracting factor" of the distractor works to disperse test takers' selection of correct answers so as to know their overall performance in answering. The purpose of doing so is not only to understand the situation where test takers who performed well overall in the exam, that is, those with higher overall scores, choosing the correct answers. Similarly, we hope to understand the situations where test takers who performed poorly overall in the exam, i.e., those with lower overall scores, chose distractors and correct answers. The RASCH-GZ system we developed and upgraded has achieved the above technical requirements in the IA process.

There are various methods for calculating the discrimination coefficient. For large-scale exams, the bi-serial correlation coefficient (R_{bis}) is generally used. We calculate the correlation between the total score of the test takers and their correct answers to the test using (R_{bis}). The RASCH-GZ we developed and upgraded uses (R_{bis}) to analyze and process the data. The RASCH-GZ system not only achieves the determination of the correlation coefficient for the correct answer of each question, but also calculates the correlation coefficient for each distractor of each MCQ. This provides a scientific basis for quality control, testing, and modification of test questions, and contributes to the current large-scale exams. The formula for calculating (R_{bis}) is as follows:

$$r_{bis} = \frac{M_r - M_w}{St} * \frac{p(1-p)}{y} \tag{2.11}$$

In the above equation:

Mr = the average score of the total score from the test takers who chose the right answers

Mw = the average score of the total score from the test takers who selected wrong answers

P = the ratio of test takers who choose the right answer to all test takers

Y = the vertical line that distinguishes P from (1-P) in a normal distribution
St = SD of the total scores based on all the test takers

2.1.14 Interpretations of Discrimination

In CTT, discriminations can be subject to various interpretations. Popham (as cited in Lynch, 2003) suggests that a discrimination of 0.4 and higher is a sign of a "very good item", and that anything less than 0.2 is a sign of a "poor" item. Items with discrimination indices between 0.2 and 0.4 can probably be improved through revision. Gui Shichun (1986) believes that a discrimination index of 0.3 or above for the correct answer of a MCQ type is acceptable. However, when using (R_{bis}), there exist at least two limitations: (1) one is that such values have no standard estimation error; (2) the other, unlike r-, is limited to the range within -1 to +1. We took these two points into consideration during the design and development of the Rasch-GZ system. As there exists no a priori specific value that CTT provides that indicates this ideal magnitude, we managed to keep the range of discrimination values between -1 and +1 just like r. On this basis, the significance of setting discrimination values for the Rasch GZ system can be attributed to the following four aspects.

2.1.15 Discrimination for Correct Answers

There are several cases for us to consider a correct answer of MCQ type. At the first place, the discrimination for correct answer cannot be less than 0.3 (which indicates a weak item, requiring professionals to moderate or to abandon it). Discrimination below 0.3 will be taken as poorly designed item and rejected. Next, the discrimination for correct answer equal to or greater than 0.3 and less than 0.7 is fully accepted, indicating a good item. The discrimination at or greater than 0.7 is a very good item but difficult to obtain in large-scale examinations. Correct answer items with high discrimination index in this case requires both professional design and moderation based on pre-test data. Third, if the discrimination index for a correct answer is 1, which is, of course, the perfect index, indicating that all test takers who scored high in the exam answered the question correctly. In contrast, all test takers with lower scores answered this question incorrectly. But this is impossible in any large-scale exam! This can only be used as a concept for classroom teaching. Therefore, consistent with the above cases, it is taken in CTT in general, that the greater the discrimination of an item, the better.

2.1.16 Discrimination for Distractors

At this very outset, we should emphasize that any distractor used in MCQ type must produce some distracting effect. The key point here is to what extent that we should maintain the distracting effect. Therefore, the following three points are worth mentioning: (1) The discrimination of a distractor equal to or lower than 0.1 would be considered a poor item, requiring professional moderation according to the context involving the stem, the key item and the other distractors, or simply abandoned. (2) Distractors with discrimination being to or less than 0.2 but greater than −0.2 can be accepted. (3) Distractors with discrimination greater than 0.3 would be taken as too strong and need to be moderated appropriately.[2]

2.1.17 Zero Discrimination

Regardless of the correct answer or a distractor, if its discrimination is zero, it means such an item does not work at all; therefore, it should be completely moderated or abandoned. The reason is simple: if it is a key, it implies no discriminating effect; if it is a distractor, it does not produce any distracting role.

2.1.18 Negative Discrimination

As the index of discrimination is a correlation, it cannot have a value less than −1.0 and greater than +1.0. However, it also follows from the reasoning above regarding the expected relationship between scores on an item and scores on the test that the discrimination of each is expected to be greater than 0.

In CTT, negative discrimination of a particular test item indicates that the test takers with higher overall scores answered the question incorrectly, while those with lower overall scores answered the question correctly. This happens in practice. For such a situation, we should first check to see whether the key to the question is entered correctly in the data file, and then check the wording of the original test item stem and all the distractors. This is usually due to improper wording by the test item writer while designing the question stem or/and distractors. In IA, for test items with negative discrimination based on the pre-test data, the common practice is to abandon it.

2.1.19 The Relationship between Item Discrimination and Item Difficulty

In CTT, test items that are either too difficult or too easy have poor discrimination, while the ones with intermediate difficulty have better discrimination. In addition, discriminability is also one of the methods for measuring test reliability. Improving discrimination is one of the ways to enhance the reliability of testing.

In the practice of large-scale exams, great importance must be attached to the relationship between item discrimination and item difficulty. The RASCH-GZ system prioritizes the item discrimination. To be more exact, accepting or discarding a test item after a pre-test largely depends on its discriminability. For example, in CTT, a test item with correct answer rate of 2% is considered a very difficult test item. However, if its discrimination is very low (<0.2). In this case, this item is usually recommended to discard or to undergo moderation before being used in another pre-test.

2.2 Standardized Examination Defined

What is standardized examination? One thing is certain that the term "standardized examination" is not available in any general English dictionary. According to W. James Popham (2005), Ruth Mitchell (2006), and Scott Jaschik (2011), standardized tests are defined as "any test that's administered, scored, and interpreted in a standard, predetermined manner". The tests often have MCQ that can be quickly graded by automated test scoring machines. Some tests also incorporate essay writing or open-ended questions that require human grading. Similar to Gui Shichuan (1986), *standardized exams are mostly large-scale, norm-referenced tests, usually using separate and objective test item type that emphasize both ability and speed.* In the author's opinion, the above only defines standardized exams based on their nature, but in terms of establishing standardized exams and of strictly controlling errors in specific, it is far from simple.

2.2.1 High-stakes Examination Defined

The term "high-stakes" exam has appeared after the practice of standardized exams. Madaus (as cited in Dianne Wall, 2005) defined high-stakes examinations as those whose results are seen—rightly or wrongly—by students, teachers, administrators, parents, or the general public, as being used to make important decisions that immediately and directly affect them. It is generally accepted that high-stakes examinations and standardized examinations are closely related with the former being a specific form of the latter, both of which carry the same crucial consequences for test takers globally, i.e., high scores on these exams may bring a series of good outcomes such as students can progress to the next grade level, mostly graduate and enter universities for higher learning, teachers get promoted, and schools obtain continued financial supports from various channels. Conversely, low scores would lead to a series of bad consequences or negative repercussions, so to speak, such as preventing student advancement, teachers' termination or even school closures. Major high-risk exams are often difficult, highly competitive, and much pressure on candidates. Typically, the Scholastic Assessment Test (SAT) and American College Testing (ACT) in the United States, the National College Entrance Examination (GaoKao) in China, and the A-Level exam in the United Kingdom. In addition, there are language PTs (of high stakes) that non-native English speakers need to pass when applying to universities or immigration programs in English-speaking countries, such as the well-known TOEFL and IELTS. What characterizes these language exams can be briefly summarized in two major aspects: Firstly, these exams cover a wide range of topics and require candidates to have a solid knowledge foundation and good test-taking skills. Candidates are under great pressure and may face both psychological challenges. Next, candidates usually invest a lot of time and resources in preparation to ensure that they can achieve their desired results. The reason is simple. Such exam results directly affect candidates' college admission, vocational qualification certification, or career development.

2.2.2 Advantages and Disadvantages of High-stakes Examination

High-stakes examinations have both advantages and disadvantages. In the author's opinion, though they have been particularly controversial, one thing is certain, i.e., they are leading to the practice of "teaching to the test", and having positive impacts on teaching especially in China. Popham (as cited in Dianne Wall, 2005) believed that high-stakes testing could lead to educational improvement. Popham used the term "measurement-driven instruction" (MDI) to describe this phenomenon. According to Popham, high-stakes tests have two different consequences for students and for both teachers and schools. He argued that if such tests are "properly conceived and implemented focusing teaching on what they assess is sensible. He rejected the notion that MDI leads to curricular reductionism, curricular stagnation, constrained teacher creativity, and lowered student aspirations. Apart from this, he supported his MDI by citing studies about "substantial improvement" in the number of students reaching established standards on basic skills tests when MDI is properly practiced (Dianne Wall, 2005).

Table 2.2 shows the Advantages and Disadvantages of High-stakes Examinations.[3]

TABLE 2.2

Advantages and Disadvantages of High-stakes Examinations

Advantages	Disadvantages
• Academic Rigor and Achievement; • Promote academic rigor; • Provide evidence of academic mastery; • Beneficial for students' success in college and careers; • Competencies thus evaluated essential for various educational and professional paths; • Correlation with Positive Outcomes.	• Teaching to the Test; • The pressure to perform well on high-stakes tests; • Lead schools to lower standards and focus primarily on test preparation; • Erode the quality of teaching and limit students' learning to test-related content; • Low scores can be detrimental to educational quality and equity.

Overall, while high-stakes examinations can drive academic achievement and correlate with positive life outcomes, they also pose challenges by potentially narrowing educational focus and imposing severe consequences for underperformance (Ebel, 1966, as cited in Kirkland, 1971; Eckstein and Noah, 1993).

2.3 Early History of Standardized Examination in China

The earliest known standardized tests can be traced back to Imperial China in the 7th century. The tests, based on the rigid "eight-legged essay" format, tested the applicants' rote-learned knowledge of Confucian philosophy. After the founding of the People's Republic of China in 1949, the former Soviet Union's five point-scale scoring system (Ge Tianmin, 1950; Sun Jie, 2014) was adopted, and standardized exams were once labeled as capitalist and neglected. Among the criticism and misunderstanding of standardized exams at that time, one point that cannot be ignored is quality problems with the exam itself, which requires exam researchers and developers to strengthen research and continuously improve exam techniques. However, during the Culture Revolution (1966–1976), all the examinations, regardless of standardized or not, were totally abandoned across the country. In 1978, China resumed GaoKao[4] and standardized exams were officially implemented nationwide. The annual number of candidates taking GaoKao skyrocketed to approximately 10 million a year. In fact, the examination is ranked the first important yet competitive examination with high stakes not only to test takers but also to parents, families, teachers and schools and beyond across the country.

2.4 Pros and Cons of Standardized Examination in China

Although standardized examinations have become an important part of China's education since 1985, problems remain. If the problems are mainly politically oriented in the past, the problems today are of technical-prone. There have been existing controversial issues for its measurement attributes, validity studies, washback effects, and systematic practice (Gui, 1986, 1995; Yang and Gui, 2015; Bachman, 2010). Take it from the author's experience as a member of MET equating project for ten years (1989–1999), the pros and cons of standardized examination in China can be briefly listed in Table 2.3.

So, that's the situation in China. This is a big debate that cannot be fully addressed in one chapter. The author only makes a brief exploration based

TABLE 2.3

The Pros and Cons of Standardized Examinations in China

PROS	CONS
Pro 1: Standardized tests are based on modern testing theory, giving objective and fair measurement.	Con 1: Standardized tests are the foreign rubbish[5].
Pro 2: Standardized tests are useful metrics for both teacher and school evaluations.	Con 2: Standardize tests are eroding the quality of teaching to "teach to GaoKao" in class and encouraging teachers to "make money outside campus".
Pro 3: Standardized tests give equal chance to all the students to compete for success.	Con 3: Standardized tests are beneficial to students in big cities only.
Pro 4: Standardized test scores indicate potential for future success.	Con 4: Standardized test scores are by no means the predictors of future success.

on personal experience of both conducting MET equating and participating in other standardized exams in China.

2.5 The Arguments of Pros vs. Cons at a Glance

2.5.1 Pro 1 vs. Con 1

Pro 1: Standardized tests are based on modern testing theory, giving an objective and fair measurement.

(1) MET equating project group and administers explain that measures are taken to ensure test score fairness by applying IRT and Rasch model to conduct test equating since 1989. Data annually obtained via sampling and equating are particularly important to help produce re-scaled scores to ensure test fairness across populations of test takers because they are the only available, consistent, and objective source of data about score reporting, and further exploration and practice would fix the deep-seated problems inherent in standardized examinations in China.

(2) Is this a fact? Isn't the TOEFL, GRE, GMAT, SAT, and ACT still administered in the United States every year?! Hong Kong test was launched in 1976 and was administered till 1986 and was cancelled purely for administrative arrangement.

(3) The first edition of "Educational Measurement" in the United States was published in 1951, and two revisions were made in 1971 and 1989, respectively (mostly rewriting and adding). Linn (1989) reviewed the

changes and developments over the past few decades in the third edition, stating that "seemingly identical measurement procedures can conceal significant changes, such as comparing the question sets of standardized achievement exams used in recent years. We can see that the use of multiple-choice questions, equivalent point scoring, and percentile ranking remains very consistent".

Con 1: Standardized tests are the foreign rubbish.

(1) The ongoing practice of standardized tests is "foreign rubbish". The United States was the first to implement "standardized examinations and then Western European countries followed suit". However, after a few years of novelty, the drawbacks of its rigid dogma that stifles spirituality were exposed. Therefore, in 1986, a hundred professors and scholars jointly petitioned the Federal Government to ban this stuff, and the US government issued an order to ban it the following year. "Standardization" became garbage.

(2) The Education Department of former Hong Kong Special Administrative Region also introduced "standardized examinations" in 1985. However, it stopped using it within a year, probably because they may have smelled the smell of garbage.

2.5.2 Pro 2 vs. Con 2

Pro 2: Standardized tests are useful metrics for both teacher and school evaluations.

(1) This is the core of standardized examinations. Standardized exams are designed to be objective measures. They assess students based on a unified test paper consisting of the same set of test items which is administered under almost the same testing conditions and during the same time periods. MCQ parts are graded by optical machine, and essay writing parts are blind reviewed. They provide an accurate, unfiltered measure of what students have actually learnt/taught. Therefore, the scores thus obtained are fair and comparable not only within the same school but also with their peers in other schools. In this way, we could determine what we're doing well within our educational continuum and where we need to improve and to invest.

(2) We view standardized testing data not only as another set of data to assess student performance, but also as a means to help teacher evaluations so as to reflect on the school curriculum for the following years. In a word, standardized tests provide a consistent measure across classrooms and schools.

(3) What is more, the more data we are provided, the more accurate the teacher evaluation decisions will end up being. Thus, standardized examinations are useful metrics for both teacher and school evaluation.

Con 2: Standardized tests are eroding the quality of teaching to "teach to GaoKao" in class and encouraging teachers to "make money outside campus".

No!

Standardized tests are "one-size-fits-all tests" which greatly discourage students who scored lower; therefore, the tests themselves become the focus of education. Teachers are not using textbooks. Instead, they are writing, copying, downloading test items or test papers, and administering the self-made simulated GaoKao test papers in class. After school, they run coaching class to show students how to deal with objective test items and guessing factors. In doing so, the testing skills may be improved but not the overall level of knowledge used to be taught in class.

2.5.3 Pro 3 vs. Con 3

Pro 3: Standardized tests give equal chance to all the students to compete for success.

Standardized tests give equal chance to all the students to compete for success. Whether by marginalized groups, remote areas, minority, disability, or other differences, students can register to take part in the standardized examination. And the test is a unified test paper consisting of the same set of test items administered under almost the same testing conditions and during the unified time period across the country. And the scores are graded using the same rubrics. For example, MCQ parts are graded by optical machine and essay writing parts are blind reviewed. All the test takers have the equal rights to check via phones, or emails, their scores.

Evidences show that every year, students who scored higher entered best universities and colleges in local provinces or in Beijing are not from cosmopolitan cities.

Con 3: Standardized tests are beneficial to students in big cities only.

No!

Standardized tests are beneficial to students in big cities only! Students who are legal residents of big cities and who scored lower may enter good universities. And it is unfair that just by luck of birth that a kid born in a big city is somehow entitled to good higher education. That's why we have so many GaoKao immigrants from marginalized or remote areas to big cities in one way or another each year.

Another reason is that well-resourced students in big cities have far greater access to test preparation, tutoring and even taking the test multiple times. In contrast, such opportunities are not afforded by the less affluent families or adults in the poverty-stricken areas .... In this sense, standardized examinations are, to some extent, the measures of students' family background and economic status rather than of their ability to succeed. Given the same conditions, all can succeed.

2.5.4 Pro 4 vs. Con 4

Pro 4: Standardized test scores indicate potential for future success. Standardized test scores are correlated with better universities or colleges. For example, MET scores help students, teachers, and parents understand which students are on track and which are falling behind so that they can act accordingly to better meet students' needs.

Furthermore, standardized tests provide evidence of academic rigor. Graduates from better colleges and universities are correlated not only with educational attainment and job market outcomes but also with going abroad for pursuing further studies.

Con 4: Standardized test scores are by no means the predictors of future success.

(1) At best, standardized tests can only evaluate rote knowledge of math, poems, and English grammar. The MCQs cannot evaluate knowledge areas such as creativity, problem solving, critical thinking, artistic ability, and etc.

(2) Another point is that the biased writing topic or test items seriously affect the quality of grading. One example is a GaoKao essay writing task in a Chinese test. It asked about "Bamboo Growing Process". This is the typically biased topic because the test takers in the southern part of China were all familiar with the subject of bamboo, and the students in the Northern part of China had no idea about bamboo growing. Many have never even seen with their own eyes what bamboo looks like.

(3) Furthermore, not all graduates from prestigious universities have found good jobs in society, and there are many who are at a loss in their careers and scientific research. On the contrary, those who did not have the chance to attend university can still make achievements in their careers and business development.

Therefore, test-skeptical parents and others say that standardized test scores are by no means the predictors of future success.

Frankly speaking, such pros and cons were obviously split between professionals and non-professionals, or between researchers and teachers or between public media and academic papers. And in specific, the gravity of the results can be narrowed down to two points: One is the concept got confused. That is, standardized tests were simply thought of as MCQs; test items were conflated with test papers. Little was known about the testing theory inherent in the practice of standardized tests. According to Gui (2017), people often confuse these two concepts: Test items and test paper. Are the "test items used in standardized exams" "standardized items"? A standardized examination establishes a unified standard based on the norm as reference. This is an examination procedure. And the test items are basically the means used to establish such a unified standard. As for test items, there is no distinction between what is standardized and what is not. It cannot be said that all the test items used in standardized exams are *"standardized"*, nor can it be said that should there be any poorly designed test item(s), the entire standardized examination will be problematic. The other is the serious scoring error inherent in subjective questions. And this is largely due to intra-rater and inter-rater inconsistency which is a universal issue, not confined within the practice of standardized examination in China. Needless to say, such misconceptions were quickly clarified.

2.6 The Establishment of Norm

The final step in standardized examination is to establish norms. As mentioned in the proceeding section, the raw score obtained from an exam is almost meaningless unless it is supplemented with additional information, as Angoff (1984) put it, that will place the score in an appropriate interpretive context. There are two kinds of meaning/information that are attached to the term norms. One is associated with notions of acceptable, desired, or required standards. The other is the statistical meaning in terms of which educational and psychological measurements are most often interpreted (Angoff, 1984). We can say that one test taker scored 85 points, which is higher than 90% of all the test takers sitting the same test; We can also say that one candidate scored 85 points, but among all candidates, this score is only higher than 50% of all the test takers. In this context, even with the score of 85, the meaning it conveys is different. Such supplementary information is actually the norm we are addressing. In CTT, norms help us understand the meaning of fractions. According to Gui (1986), a score can often be compared among different groups of candidates, for example, when comparing it among candidates nationwide, we use the national norm; comparing it among candidates across the province, we are using the provincial norm; comparing it among all the candidates in the school, we are using a school norm. The norm used depends on the purpose of the exam. Taking earlier MET in China as an example, it was used to select students to enter universities throughout Guangdong province, and the provincial norm should be used. Taking WSK in China as another example, it was used to determine the English proficiency of test takers nationwide studying abroad, and the national norm should be used.

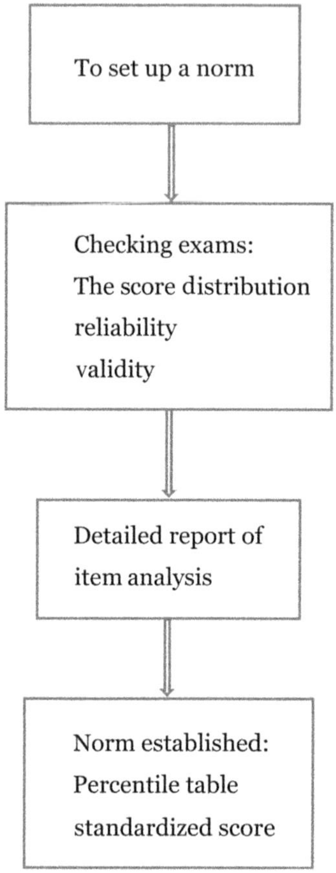

FIGURE 2.5
The flowchart for establishing the norm

As discussed above, exams are a means of establishing norms, and before establishing norms, quality checks must be conducted on the exams. Let's give a simple analogy. If we want to use a ruler to measure someone's height, we need to first prove that our ruler is up to standard. For better understanding, the flowchart for establishing the norm is presented in Figure 2.5.

2.7 Summary

This chapter focuses on the introduction to the basic concepts regarding IA of CTT and their relevant calculation methods involved in the RASCH-GZ system. These are very important concepts for language testing research.

They are: correct answer rate, P-value, item difficulty, skewness, kurtosis, range, reliability, α-value, Δ-value, item discrimination as well as discrimination analysis and explanation for distractors in MCQ type. It is worth mentioning that all the calculations discussed herein are actually realized via the Rasch GZ system developed by the author. For ordinary users, we only need to understand the ideas of these concepts as a basis to check the quality of the test items we design so as to decide whether to accept, to moderate or to discard the relevant (calibrated) items. The Rasch GZ users can obtain detailed IA of each test item with each option as well as detailed test reports by simply moving their fingers or clicking the mouse at the Rasch-GZ operating menu. The detailed IA and the test reports are helpful documents or scientific guidelines, so to speak, for Chinese teachers of English, graduate students, staff members of various examination authorities, both testing professionals and practitioners to collect and analyze examination data, formulate and moderate test items, conduct test equating for large-scale examinations, and eventually establish item bank for CAT. As for the specific operation of the Rasch-GZ system, readers can refer to Chapter 4 of this book.

Apart from this, the practice of standardized examinations in China is addressed in some details, tracing back to the test practice in the earlier days in China, the controversy over the standardized examinations and solutions in the recent years. Finally, the establishment of norm for standardized examinations is discussed. It is the author's hope that the present chapter could silhouette the whole picture of IA of CTT and the retrospective and status quo of standardized examinations in China.

Notes

1. WSP, abbreviated from Chinese Characters "Waiyu Shuiping Kaoshi", literally meaning Foreign Language Proficiency Test" refers to the English Proficiency Test for Chinese people to go abroad for visiting scholar, research scholar or other purposes with financial supports from government channels. The predecessor of WSK was EPT (i.e., English Proficiency Test), later changed to PETS (i.e., The Public English Test System) which is a(n) (open) national English proficiency testing system designed and implemented by the Education Examination Authority of the Ministry of Education (formerly the Examination Center of the Ministry of Education). PETS candidates, regardless of their age, educational background, registered residence and other backgrounds (but due to policy reasons, not for students who are still receiving nine-year compulsory education), can choose their own level to register so long as they have learnt a little English. In December 2016, the Anhui Provincial Education Enrollment Examination issued a notice clarifying that the PETS exam would be suspended in Anhui Province in 2017, becoming the first province in China to stop the PETS exam. Subsequently, since 2016, several provinces including Anhui Province, Hebei Province, Hubei

Province, Jilin Province, Chongqing City, Xinjiang Uygur Autonomous Region, Shanxi Province, Shandong Province, Jiangxi Province, Henan Province, Sichuan Province, and Fujian Province have successively issued notices to suspend exams.

2. There is also some different voice (Yigal Attali and Tamar Fraenkel, 2000), showing that using the point biserial as a discrimination index for distractor and examinees who did not choose the distractor is theoretically wrong and may lead to an incorrect rejection of item. The author will discuss this point in separate paper.
3. Due to space inconvenient digression, the author will discuss the impact of high-risk exams on teaching in separate article. This section only summarizes the advantages and disadvantages of high-risk examinations in general.
4. To be more exact, "GaoKao" is Chinese, pronounced as [gāokào], referring to university entrance examinations which is administered by China's Ministry of Education once a year across China. GaoKao is mandatory for middle school students who want to enter any institute of higher education; therefore, GaoKao scores determine young people's career trajectory. In other words, GaoKao scores carry important consequences for students, teachers, and schools: a poor performance on GaoKao means completely out of question to enter a university. Therefore, those who failed for the first year would mostly apply for re-attending the courses of the last year of high school so as to get better review and preparations to re-take GaoKao next year. That's why the number of GaoKao-takers is so big in China. Take the year of 2006s for example, it reached 9.5 million. Another term closely associated with GaoKao is MET abbreviated from Matriculation English Test consisting of 85% multiple choice question type and 15% writing task. MET is one important English entrance examination of GaoKao. In this book, these two terms may be used alternatively according to the context.
5. At that time, there were prominent headlines in newspapers such as "Standardized Examination and Foreign Garbage", *Southern Daily*, the 25th of April, 1999. For quite some time, the reporting made a certain negative impact in China because in terms of newspaper circulation, several academic papers could not compete with it.

Bibliography

Alderson, J.C. (2011). A Lifetime of Language Testing. Cambridge, UK: Cambridge University Press.

Angoff, W.H. (1984). Scales, Norms, and Equivalent Scores. Princeton, NJ, USA: Educational Testing Service.

Attali, Y., & Fraenkel, T. (2000). The point-biserial as a discrimination index for distractors in multiple-choice items: Deficiencies in usage and an alternative. Journal of Educational Measurement, 37(1), 77–86. https://doi.org/10.1111/j.1745-3984.2000.tb01078.x

Bachman, L.F. (2004). Statistical Analyses for Language Assessment. Cambridge, UK: Cambridge University Press.

Bachman, L.F., & Palmer, Adrian S. (1996). Language Testing in Practice. Oxford, UK: Oxford University Press.

Brown, J.D. (1988). Understanding Research in Second Language Learning: A Teacher's Guide in Statistics and Research Design. Cambridge, UK: Cambridge University Press.

Brown, J.D. (2001). Using Survey in Language Programs. Cambridge, UK: Cambridge University Press. ISBN-10: 0521796569

Crozier, Justin (2002). "A Unique Experiment," China in Focus.

Davies, A., Brown, A., Elder, C., Hill, K., Lumley, T., & McNamara, T. (1999). Dictionary of Language Testing. Cambridge, UK: University of Cambridge Local Examinations Syndicate and Cambridge University Press.

Ebel, R.L. (1966). The Social Consequences of Educational Testing. In A. Anatasi (Ed.), Testing Problems in Perspective (pp. 18–28). Washington, D.C.: American Council on Education.

Eckstein, M.A., & Noah, H.J. (1993). Secondary School Examinations: International Perspectives on Policies and Practice. New Haven, CT, USA: Yale University Press. ISBN-10: 0300056411

Fair Test (2000). "Florida Teachers Refuse Bonuses for High Test Scores," fairtest.org, Jan.

Hatch, E., & Farhady, H. (1982). Research Design and Statistics for Applied Linguistics. Rowley, MA, USA: Newbury House Publishers. In The Research Manual: Design and Statistics for Applied Linguistics. Massachusetts, USA: Newbury House Publishers, Inc., Rowley.

Hatch, E., & Lazaraton, A. (1991). The Research Manual: Design and Statistics for Applied Linguistics. Massachusetts, USA: Newbury House.

Hughes, Arthur (1989). Testing for Language Teachers. Cambridge, UK: Cambridge University Press.

Jaschik, Scott (2011). "Can You Trust Automated Grading?," insidehighered.com, Feb. 21, 2011.

Kirkland, M.C. (1971). The effects of tests on students and schools. Review of Educational Research, 41(4), 303–350.

Kolodner, Meredith. (2011). "Students, Teachers Sweating High-Stakes Tests as Parents Rebel against Constant Prep," *Daily News*, May 3, 2011.

Linn, R. (1989). Current Perspectives and Future Directions. In R. Linn (Ed.), Educational Measurement (3rd Ed [C]). Washington, DC, USA: American Council on Education/Macmillan Publishing Company.

Livingston, Samuel A. (2004). Equating Test Scores (without IRT). Princeton, NJ, USA: Educational Testing Service.

Lynch, Brian K. (2003). Language Assessment and Programme Evaluation. In Alan Davies & Keith Michal (Eds.), Edinburgh Textbooks in Applied Linguistics. Edinburgh University Press. Available at http://www.euppublishing.com

Madaus, G.F. (1988). The Influence of Testing on the Curriculum. In L.N. Tanner (Ed.), Critical Issues in Curriculum: Eighty-seventh Yearbook of the National Society for the Study of Education 83-121, Chicago: NSSE, University of Chicago Press.

Mathews, Jay (2006). "Just Whose Idea Was All This Testing?", *Washington Post*, Nov. 14, 2006.

Mitchell, Ruth (2006). "A Guide to Standardized Testing: The Nature of Assessment," centerforpubliceducation.org, Feb. 15, 2006

Morin, Amanda (2011). "What Is High Stakes Testing?" about.com (accessed June 20, 2011).

Popham, W.J. (1978). Criterion-referenced Measurement. Englewood Cliffs, NJ: Prentice Hall.

Popham, W.J. (1990). Modern Educational Measurement (2nd Ed). Englewood Cliffs, NJ: Prentice Hall.
Popham, W.J. (1994). The instructional consequences of criterion-referenced clarity. Educational Measurement: Issues and Practice, 13, 15–18, 30.
Popham, W.J. (1997). What's wrong- and what's right with rubrics. Educational Leadership, 55(2). Retrieved online, May 2002: http://www.ascd.org.reading-room/edlead/9710/popham.html
Popham, W.J. (2005). "Standardized Testing Fails the Exam," edutopia.org, Mar. 23, 2005
Thorndike, R.L. (1971). Educational Measurement (2nd Ed). Washington, D.C.: American Council on Education.
Wall, Dianne. (2005). The Impact of High-stakes Examinations on Classroom Teaching: A Case Study Using Insight from Testing and Innovation Theory. Cambridge, UK: Cambridge University Press. ISBN: 9780521542494
桂诗春. (2017). 标准化考试"辩. In Gui ShiChun (Ed.), Selected Academic Papers of Gui Shichun. Shanghai Foreign Language Education Press.
桂诗春. (1986). 《标准化考试:理论、原则与方法》. [M]. 广东高等教育出版社.统一书号: 7343•20
桂诗春. (1995). 对标准化考试的一些反思 [J] 中国考试 (3) 。
张厚粲. (1986). 《心理与教育统计学》. 北京师范大学出版社.ISBN 7-303-00039-9/G-36
李绍山. (1999). 《语言研究中的统计学》. 西安交通大学出版社.ISBN 7-5605-1118-X
葛天民. (1950) 介绍苏联五级分制记分法。来自掌桥科研 DOI: CNKI: SUN:RMJY.0. 1950-04-014
孙杰. (2014) 新中国中学生成绩评定方法的实验—以五级分制记分法为个案的研究. 《CNKI》

3

Rasch Model and Use Cases

This chapter introduces the Rasch-based use cases in Pacific-Rim countries and regions with focus on research background of test equating and the specific problem solving in three large-scale yet of higher stake tests: MET, College English Test (CET), and Chinese Proficiency Test (HSK) in China Mainland. The basic concepts related to test equating, the significance, the conditions under which test equating to be conducted, and the specific procedure of test equating are demonstrated with examples. Apart from this, based on the personal experience, the author listed with specific details all the major IRT-based and Rasch-based computer software used for test equating home and abroad. The understanding of item parameters and ability estimates after test equating is also addressed. For more in-depth theoretical discussions on test equating, readers can refer to the authoritative reference list in both Chinese and English provided at the end of this chapter.

3.1 The Status Quo of Rasch Measurement in Hong Kong, SAR, China

The use of the Rasch Model was first established at the Hong Kong Examinations and Assessment Authority (HKEAA) and the Education University of Hong Kong (EdUHK) in 2007. The HKEAA is responsible for the administration of standardized public examinations. The authority administers the Hong Kong Diploma of Secondary Education Examination (HKDSE). Previously, it developed, administered, and assessed the Hong Kong Certificate of Education Examination (HKCEE) and the Hong Kong Advanced Level Examination (HKALE). For more than a decade, the HKEAA psychometric team has used the Rasch Model for developing unidimensional tests, setting benchmarks and cut-points, and maintaining those set standards across assessments and over time. Over the same time, the EdUHK has appointed world leaders in Rasch measurement such as Prof. Trevor Bond to its faculty. They implemented Rasch measurement theory to large-scale research projects and taught Rasch measurement and supervised students at the graduate level. The research center called Assessment Research Centre (ARC) at university level is a core unit committed to the research and education of Rasch measurement. One exemplar of the projects conducted

DOI: 10.1201/9781003665595-3

by ARC was the Assessment Project (2005–2008) funded by the Hong Kong Education Bureau. This project was a large-scale developmental project aiming to promote the application of Rasch measurement in primary and secondary schools with over one hundred schools involved and with more than one thousand teachers trained to use Rasch analysis to support formative assessment in their classrooms. However, at the same time, the adoption of Rasch measurement remained in its infancy across the other universities in Hong Kong (Bond, 2016).

3.2 Research Background of Test Equating in China Mainland

According to Liu, Yuming (2020), it is generally accepted that in China before 1980, **most people** within the professional field of language testing were unaware of test equating. From the early 1980s, Chinese testing professionals came to be aware of its importance. In the author's opinion, there are two crucial reasons for the continued interest in test equating in China: at the first place, the society as a whole was attaching extreme importance to the issue of fairness of the test scores obtained from large-scale yet high-stakes examinations, the scores from MET, CET, and HSK in particular. Next, technically, professionals began to realize that the ongoing CTT practice of the estimation of test item difficulty cannot solve the problem of score fairness in China. This has become a major concern for Chinese government and professionals to realize the importance of the equating scores across time, across test takers, and across test papers. In 1977, China resumed the entrance examinations for higher education (or GaoKao in Chinese, or MET, the Matriculation English Test) in all the provinces; the total number of candidates had reached as many as 5.7 million in that year. 1978 saw the implementation of a unified and standardized examination at the national level. Since the implementation of the unified examination, enrolment had once been based on raw scores. This was the crux of the problem. At that time, there were two most controversial issues: first, for such a large-scale, high-stakes examination, how to control the overall difficulty of test paper each year? Second, it was unscientific to select the "best students" on the basis of raw scores. In other words, it is totally unscientific to simply add up the raw scores (i.e., the scores graded on each of the candidates' test papers) of different subject because the difficulty of test papers varies from year to year (Gui Shichun, 2017). These two issues must be addressed immediately; otherwise, the practice of standardized tests would not be feasible in China. From a more professional point of view, these problems are actually the problems of seeking solution to equate the difficulty of test items in different test papers administered to different test takers in different years. Therefore, the calibration of test item difficulty and the implementation of test equating were officially put on the agenda. In 1987, the Examination Centre directly under the Chinese Ministry of Education was officially established. From 1990 to 1999, with the strong support of the Examination Centre, a 5-member team was set up, with Prof. Gui Shichun as its professional head, successfully implemented the MET Equating Project[1] for a period of 10 years. Due to space, to make a long story short here, since the implementation of the GaoKao or MET equating project, there have been a total of six large-scale, high-stakes standardized exams that have been implemented across China, all of which have an international impact in one way or another. They are the Gao Kao English (MET) since 1977 (Gui et al., 1993), the CET since 1987 (Zhu et al., 2003), the Hanyu Shuiping Kaoshi (HSK) since 1991 (Xie, 2022, 1998), and the General English Proficiency Test (GEPT) in Taiwan Province since 2000 (Rachel Yi-fen Wu, 2020), the Diploma of Secondary Education Examination (DSE) in China Hong Kong since 2013 (Tian, 2006), and the Joint Admission Examination for Four Higher Education Institutions in Macao (JAEfFHEI), which is in the pipeline since 2017. Moreover, all these examers apply the Rasch model to achieve test equating, and thus have higher reliability and validity. Among them, MET is the one whose number of participants from China Mainland is the largest in the world. As early as in 2006, it had reached 9.5 million. The author conducted a literature search in China National Knowledge Information Network (CNKI)[2] with the keywords in Chinese "高考入学考试" (college entrance examination). The results are shown in Figure 3.1.

In Figure 3.1, the number of students taking MET from 1977 to 2020 (excluding China Hong Kong, China Macao, and Taiwan Provinces) is shown, where green indicates the annual total, orange, the actual enrolment, and the red dotted line, the annual enrolment rate.

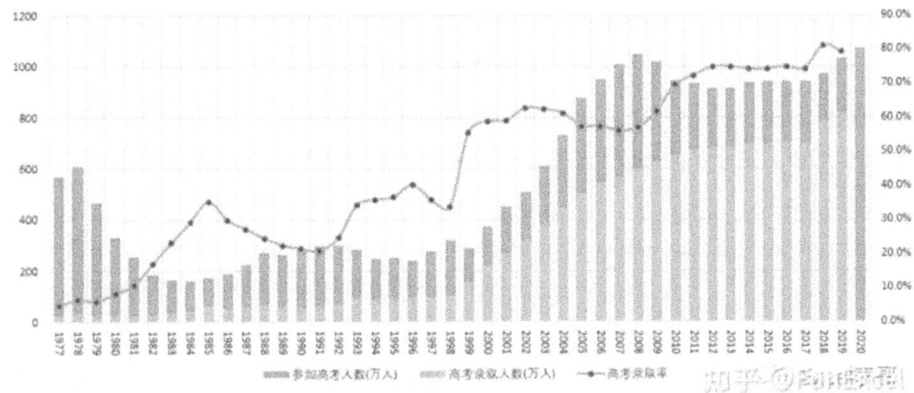

FIGURE 3.1
Summary of the number of candidates who took the college entrance examination in the Mainland China, 1977–2020.

3.2.1 Test Equating Defined

In the language testing community, equating has traditionally been defined in different ways. According to Liu, Yuming (2020), in the terminology of educational and psychological testing standards, equating is more accurately described as "obtaining comparability by means of scales"; According to Holland and Dorans (2006), Linn (1993), and Mislevy (1992), equating is the linking or matching of different test papers. To quote Kolen and Brenman (1995), linking of test forms through common items so that scores derived from the tests which were administered separately to different test takers on different occasions after conversion will be comparable on the same scale. Then let's trace back to Rasch (as cited in Andrich and Ida Marais, 2019). Often, when two sets of items with some common items are placed on the same scale, it is said that the sets of items have been linked. Such a design was in the original work of Rasch which led him to his theory or measurement. Apart from this, what impressed the author most is the further explanation of the idea of test equating given by Kolen and Brennan (2014) and translated by Liu, Yuming (2010). Below is the further explanation of the concept of test equating in two different yet representative cases.

Suppose a test taker took college entrance exams two times. The score obtained in the first test is higher than the one in the second report. One explanation is that the difference between these two scores reflects the test taker's higher level of achievement in the second test compared with the one in the first. However, if the test items in the two tests for this test taker are the same, the difference in score would not be due to the difference in the achievement level, but rather the score in the second test has been exaggerated. The reason is simple: All the test items were exposed. Fortunately, university entrance examination uses a completely new test paper each year, so it is unlikely that candidates will see the same test items in two separate tests (Liu, 2010).

Using different test papers at different testing times may also lead to another issue. Assuming two students apply for scholarships from the same university, part of the eligibility criteria for this scholarship is test scores. These two students took a test at different times. The score obtained by Candidate A is higher than the score obtained by Candidate B. One explanation is that A has a higher level of achievement than that of B. However, if the test items taken by A are easier than those taken by B, then this test is unfair to B. Because in such a situation, the difference in test scores is not the difference in the achievement level, but the difference in the difficulty of the test paper. To avoid such a problem, most university entrance tests use the method of score equating. If the scores of the test paper are successfully equalized, the difference in scores between A and B after equating cannot be attributed to the difference in difficulty of the test paper (Liu, 2010).

These two examples highlight the equating process, i.e., test equating can only be used for parallel forms of test papers and for adjusting the difference in difficulty between test papers. By equating successfully two sets of test papers, the score obtained by Candidate A on Test A can be treated as equal as the score obtained by Candidate B on Test B at different times.

Based on the above, the practice of "test equating" addressed in this book refers to a more detailed process: Linking data from parallel test papers taken by different batches of test takers of homogeneous in nature at different times and then converting the difficulty of the test items in the parallel test papers through anchor items. After equating, the difficulty parameters of the test items and the subject ability estimates are comparable on the same scale (Wright and Stone, 1979; Lord, 1980; Hambleton et al., 1985; Gui, 1985, 1989; Gui et al., 1993; Kolen and Brenman, 1995). Specifically, starting from 1989 to 1999, two MET parallel sets of test papers are connected every year through anchor items and were administered to different batches of MET candidates each year, and the data thus obtained are calibrated using GITEST to obtain the difficulty of each set of test items for equating, so that the difficulty values after equating are comparable on the same scale. In the practice of language testing, test equating is seen as a statistical process that adjusts the scores of different groups of test takers who took the parallel tests at different times and enables these scores (expressed as proficiency values) to be compared and referenced on the same scale (Kolen and Brennan, 1995).

3.2.2 Test Equating and Its Significance

Test equating is conducted to large-scale examinations with high stakes using parallel test papers that are identical in the format of the test items, the number of test items and the rubrics as in the case of MET in China, the purpose of which is to ensure the security of the test papers and the fairness of the test scores. This is because, even if the key members of test item producing team remain the same from year to year, there is no way to guarantee that there be no difference existing in the difficulty of the various sets of test papers. The comparability of test scores from test papers consisting of different sets of test items designed to test the candidates of the same ability is a crucial issue that test item writers, test experts, and test takers need to face (Hambleton and Swaminathan, 1985; Rogers, 1989; Gui Shichun et al., 1989, 2016, 2017). The GaoKao or MET in China is the most prestigious and competitive high-stakes examination, and the number of candidates now exceeds more than 10 million per year, which is a matter of concern to thousands of families and the admission standards of universities; therefore, the test item difficulty, the test security, and the test fairness must be put strictly under control, which is crucial for all the candidates and for the whole society. If the same Gaokao or MET test paper is implemented nationwide for fresh high school students in different years and used as the basis for university admission, or if schools use the same test paper for graduation criteria each year and repeat it for graduation performance assessment, then under China's current social environment, the test items will be exposed after the first implementation of the test, and it will be no way to ensure the test security and fairness to the test takers. On the other hand, it is neither feasible to administer two separate tests to the same group of students in order to compare the difficulty of parallel test papers. In this sense, the significance of a good test equating is extraordinary, which is, so to speak, the "mainstay" of language testing field today. Should the technology be guaranteed, it would greatly reinforce the educational basis in China.

As discussed above, thc significance of test equating is to ensure the security and fairness. In fact, this is a universal problem. A typical problem in China is that in 1985 when the standardized test was first implemented, a province (here we call S province) used to be ranked first in the country in terms of the scores regarding entrance examination for higher education, was not ranked the first in the following year. Instead, another province (we may call J province) whose MET scores exceeded that of S province and was ranked the first one in China. There was then a very hot dispute over the results of the standardized test at a meeting. One side was complaining, saying the test items were harder; the other side denied such an assumption, arguing: no, the test items were by no means harder and they remained unchanged. Instead, the teaching quality of S province had declined. If test equating were not conducted at that time to solve the problem at its root, it would have been impossible for China's standardized tests to go on nationwide. In this sense, the ten-year MET Equating Project (1990–1999) played a decisive role in the implementation of standardized tests in China.

In China, the entrance exams for higher education, various examinations for job promotion, school final exams, etc., are all large-scale examinations with high stakes, and each time such examinations are administered, new test items should be used. In any two parallel exams of the same kind, exposed items are not allowed to be used. Therefore, the overall difficulty of the test items in the different test papers must be different. Test equating is to adjust or to re-scale, using a Rasch-based computer software, the difficulty of the test items of the parallel tests so as to put all the equated difficulties and estimates on the same scale for comparison. Without test equating, professionals cannot tell whether there is a trend in test takers' performance over time, nor can they monitor any changes in the difficulty of test item over the years; without test equating, the overall difficulty of test items would be unstable, fluctuating from year to year, and test takers may happen to benefit from encountering easier items, while others may be disadvantaged by happening to have more difficult items. To conclude, valid test equating in language testing practice is of great importance for decision-makers and for making the test results (expressed as scores) fair and safe.

3.2.3 Two Universal Problems in Test Equating

There are two universal problems with test equating in large-scale standardized tests. According to Xie Xiaoqing (1998), through the analysis and study of HSK data, two important issues were raised that were considered to be central in the test equating adopted in HSK at that time. The first issue is the so-called "order effect" (Noguchi, 1986; Tian, 2006; Xie, 2022) caused by the design of common subject equating (CSE). According to Xie, in order to obtain reliable scores through test equating, the same group of test takers is asked to take two sets of parallel tests. For example, due to some kind of fatigue and practice effect between the two tests plus other psychological effects, test takers may not have taken the first test very seriously, or they may have just taken it as practice for the second test. As a result, although the test takers are the same, their performances and attitudes toward answering the questions during the testing period must have been different. Xie argues that such a CSE design must be problematic, and sometimes the design may go totally wrong. In addition to this, the problem of efficiency in administering/organizing examinations arises in China when two examinations are administered to the same group of candidates. Undoubtedly, whether the test results obtained from the CSE design should be regarded as reliable reference needs to be further addressed. The second central issue is the problem of the exposure of the test items in the common item equating (CIE) design. In the practice of implementing CIE, there are a certain number of linking items anchored in both the basal test and the to-be-equated test. To a large extent, linking items represent the content and overall difficulty of the entire test. They are called Anchor Items by testing professionals. Since the anchor items used for equating are repeatedly used in subsequent tests, they have become extremely greasy prey for the private training centers, tutors, and all the test takers both on and off campus across China. In case the anchor items were exposed and disseminated among the candidates both inside and outside the school, that would be serious problems not only ruin the reputation of the relevant examinations but would also greatly reduce the validity of the equating. Such a situation faced at the time was considered to be one of the main obstacles to go on with the practice of equating using linking items. Therefore, to link or not to link, that is a question!

3.2.4 Two Solutions by CET and MET

In response to these two major problems, CET solved the so-called "order effect" problem by using its own CSE method, while MET devised its own CIE method, which very well solved the thorny problem of linking item exposure.

3.2.4.1 Application of the Rasch Model to Test Equating of CET

As per Yang Huizhong (2003), to solve the so-called "order effect" problem, they used their own CSE method. For such a purpose, Yang and his research team designed a comparative test and conducted the test on a randomly sampled group of undergraduate students right at the selected observation university one week before each CET was administered. The test paper was used for the normative group. A non-significant difference between the performance of the sampled group in the comparative test and that of the normative group would indicate little change in the level of proficiency; a significant difference in the on-site CET test scores would imply a change in the overall difficulty of the test paper. Since 1990, Yang and his team have used the Rasch model to conduct score equating of CET Band-4. Since the comparison paper and the basal paper were conducted on the same group of candidates, the scores thus obtained from the two test papers could be placed on the same scale of ability and thus be comparable. In the common practice of language testing, this practice is usually what CSE is about. To quote Young (2003), in this process, an equating model is built on a random sample of candidates prior to applying the model to the entire CET population. It is vital that comparative test is conducted under quality control so that the resulting data are reliable. Since the comparative test and the field test are conducted on the same group of candidates with a time interval of merely one week, theoretically, there will be no significant change regarding the proficiency level of the candidates during this one-week period, and their English language proficiency cannot be a major factor affecting the results of the test. Furthermore, since the test items used for the CET Band-4 are designed by trained item writers under strict quality control and are well calibrated, the overall difficulty level of the test is tightly controlled; hence, there should not be a significant difference between the results of the two tests.

Excluding the above factors, there are only two underlying reasons (Xie Xiaoqing, 1998) that may lead to a significant difference between the scores obtained by the same group of subjects from the comparison test and the scores they obtained from the field test. One is the difference in terms of the candidates' attitudes toward the field test and the comparison test; the other reason is the possible exposure of the comparison test paper, as even a partial leakage of the test paper would invalidate the equating data. In view of these two factors, Young and his CET team have taken two steps since 1994 to ensure that the equating model works. The first is of psychological. Prior to the administration of a comparative test, it is announced to sampled subjects taking the test that their scores from the comparative test and the field test will be equally valid, and that subjects may choose the higher score on either test as their final exam score. To a greater extent, this approach eliminated the factor of subjects' attitudes toward test results. The next measure is of technical in nature, i.e., if the difference in a candidate's score between the comparative test and the field test (Xd) is too large, that score is considered unfit and will not be included in the equating procedure (Xie, 1998) . In this sense, the CET provides a good solution to the problems inherent in CSE method.

3.2.4.2 The Application of Rasch Model to Test Equating of MET

To solve the problem of linking item exposure, the MET equating team designed its own CIE method. By CIE is referred to as the equating method for parallel test forms with items in common. According to Mislevy and Bock (1990), CIE can be used even when the tests are administered to test takers from different populations, but requires that the two tests are congeneric, that is, measure the same underlying factor of ability. Professor Gui Shichun (1930–2017) used an anchored test randomized group design (ATRGD). Specifically, test results from MET1988 (The year 1988 was the first year of MET implementation in China) were used as basal test for equating. Following each subsequent year, test item difficulty and test taker ability were estimated using data collected from test takers from the sampled schools and connected to the basal test data. Once the difficulty of the test items is found to be harder than in those of MET1988, the relevant scores are adjusted upwards; once observed the test items are easier than those of MET1988, the scores will be adjusted downwards.[3]

As for how to solve the problem of linking item exposure traditionally used for test equating, Gui instructed his team members to take further action to implement the ATRGD in the following steps. First, the project team did not place the linking items in MET1988, nor in MET1989 and other MET test papers in subsequent years. Instead, a separate set of test papers was designed. This test paper called Equating Test Paper (ETP) identical in format with the MET test paper consists of 85 multiple-choice questions of which 28 ones were used as linking items. The second step is very important.[4] Instead of placing the 28 linking items to be used directly into the ETP in order, the project team mixed these 28 linking items with the other 57 items. The timing was chosen to allow subjects to take the ETP separately at the sampled schools three days before the actual MET was administered. In doing so, it was completely out of the question for the sampled subjects to realize, let alone "identify", which test items were used for the MET equating while taking the ETP. This measure greatly prevents any likelihood of leakage of linking items. The sampling process was carried out strictly in accordance with the invigilation requirements stipulated by the MET. The third step, which is even more important because it takes into account the psychological aspects of the sampled group, as well as the homogeneity of the sampled group in terms of age, motivation and learning background, is that the annually sampled subjects (300–400 persons) were informed that the scores of the ETP mentioned above would be regarded as the results of the school's final examination; therefore, all the sampled students treated the ETP very seriously so that the data thus collected were authentic and reliable.

Three days later, these subjects took the actual MET. Their equating data and the actual MET data were sorted out and connected via the linking items. When all the data are edited and entered into the computer, all the parameters required for equating are obtained using (GITEST) with just a single click of mouse.

3.2.5 Basic Conditions for Conducting Test Equating

To conduct test equating, the following four conditions should be met. At the first place, learn how to use a kind of test equating software. The list of computer software for conducting test equating in (3.7) is a good reference. Next, design your own equating practice according to your research needs. While the application of Rasch model to both CET and MET addressed in (3.4) serves good examples, the reference list provided at the end of section may offer more in-depth research methods. Equally important is that you need a qualified team of test item writers. The point to emphasize here is that test equating is technology, while test item writing is art (Xiaoju Li, 1997). Here, the art includes both production and moderation. The latter is, in turn, largely based on the item analysis. This is an indispensable yet important part. Without such strict production and moderation, the test equating will be a castle in the air. Finally, a good knowledge of data collection and editing is imperative.

3.2.6 Test Equating: A Simplified Example

In the language testing research, test equating is an open and general topic, differing only in the specific implementation, what model and/or which software to use. Researchers in the fields of education, psychology, statistics, language testing, and computer science have long made many great contributions in one way or another to the practice of test equating (Angoff, 1971; Holland and Rubin, 1982; Kolen and Brenman, 1995, 2004, 2020; Noguchi, 1986; Livingston, 2004; Tian, 2006; Wright and Stone, 1979; Wright et al., 1980, 2004; Wei et al., 2016). For example, apart from the Equipercentile Equating[5] used by ETS, CIE used by MET, CSE by CET discussed in Section 3.4, we also have Common Population Equating (CPE), Common Respondent Equating (CRE) (Mislevy and Bock, 1990), and even other methods different from both IRT and Rasch model such as comparative judgment (Attali et al., 2014; Baird, 2007; Benton, 2021, 2020; Bramley, 2005) so on. This shows the important role of test equating in today's large-scale yet high-stakes examinations[6] and particularly in the last decade or so, academics have paid increasing attention to the use of the Rasch model to the research of objective measurement (Zhang and Quan, 2013, 2015, 2017, 2018, 2024).

The "test equating" addressed herein is the one clearly defined in Section 3.2 of this chapter. However, the author knows merely a good definition is far from enough. Though in the Chinese context with strong examination-oriented culture, the ideas of "test equating" are still exotic to most of the teachers of English, educators, executives, management staff members, and students across the Mainland China. To ensure better understanding, the author would like to devote this section to demonstrate through illustration a simplified yet concrete example with data to further reinforce the ideas.

Let's come to the first step to understand the following statement providing a more intuitive description of what is meant by test equating:

> Test takers in Group A take Test A with L items of which n are linking items;
>
> Test takers in Group B take Test B with L items of which n are linking items.

In language testing, this has been interpreted as follows: two parallel tests connected by n linking items are administered to two different groups of examinees in the same cohort at the same or different times, and the goal to achieve is to equate all the L items in the two parallel tests and the results can be comparable on the same scale (Wright and Stone, 1979; Hambleton and Swaminathan, 1985; Gui Shichun, 1989, 2016, 2017; Zhang et al., 2000; Zhang and Quan, 2004, 2012, 2014, 2016; Zhang and Quan, 2019, 2024).

Next, to achieve this, Test A is used as a basal test, and then n items (n < L) are selected from this basal test as linking items and are placed into Test B. To be more specific, the following array shows the ideas: Items 1–16, totally 16 items, which are used as linking items. The number of test items in Test A is 42 in total; the number of test items in Test B is 42 as well.

Test A **1 2 3 4 5 6 7 8 9 10 11 12 13 14 15 16** 17 18 19 20 21 22 23 24 25 26 27 28 29 30 31 32 33 34 35 36 37 38 39 40 41 42

Test B 1 **2 3 4 5 6 7 8 9 10 11 12 13 14 15 16** 43 44 45 46 47 48 49 50 51 52 53 54 55 56 57 58 59 60 61 62 63 64 65 66 67 68

One more thing worth mentioning here: Depending on what kind of software to use, the format of the two tests usually should be the same, but the number of test items in each test may not necessarily be exactly the same.

Finally, we collect data from the test papers administered to the test takers. For example, if the two test papers above are administered to two different groups of test takers, each having 40 students. The data metrics we obtain would be edited like this:

Test Taker ID

Test A000001 **1101010101010101**01010101011010101010111

Test A000002 **0110001011010101**1110111101011010101010001

Test A000003 **1101010101010101**0101010101101010101110

… … … …

Test A000040 **1111001011010101**1111111101011010101010001

Test B000001 **1101111111110101**0101010101011010101010111

Test B000002 **0110001011110101**1110111101011010101010001

Test B000003 **1101010101010101**0100000000000001010101110

… … … …

Test B000040 **1111111011010101**11111111101011011111101001

In this dichotomously scored data metrics, the first 16 items were used linking items as described above. This is considered as a typical example of "parallel test equating". Given this, we use one of the Rasch-based or IRT-based software (Section 3.7), import the data, and the equating will be done with a single click of mouse (For details, refer to Chapters 4 and 5).

In today's testing practice, test equating plays a significant role and is regarded as a prerequisite for computerized-adaptive testing (CAT), item banking construction, Internet-based, and online testing. Through test equating, we could very well observe or monitor the variations in the overall difficulty of test items from one test to another, thus adjusting the corresponding ability estimates derived from tests in different times. Since test equating is a complex process that calibrates a large amount of data, so manual calculations are infeasible. It is in this sense that the Rasch-GZ developed by the author offers us an effective tool (see details in Section 3.7. and detailed illustration with difficulty parameters in Chapter 5).

3.3 Test Equating and Its Software Family

In the language testing community, there are two types of software that can be used for test equating. One is based on IRT such as BILOG, Logist, etc.; the other is based on the Rasch model, such as WINSTEPS, ConQuest, RUMM, etc. For a long time, there was only GITEST system based on Rasch model in China. This section gives a brief introduction to these two types of software. Interested readers can follow the references or links provided to do in-depth research or to demonstrate a whole process of test equating for classroom teaching or for large-scale examinations with high stakes.

3.3.1 BILOG

The BILOG software is considered one of the representative software based on the IRT (Linacre, 2005). Originally developed by Prof Robert J. Mislevy of the University of Maryland and R. Darrell of the University of Chicago, USA, BILOG is known as the "Item Analysis and Test Scoring with Binary Logistic Models" (Mislevy and Bock, 1986, 1990).[7]

BILOG is very powerful including CIE, CPE, and CRE (Zhang and Hu, 2000; Zhang, 2004). With the continuous updating of technology, the BILOG system now has more powerful functions; for example, apart from item calibration, test scoring, test and item information, it has the vertical equating functions, and compatible with import of WORD documents, and so on.

3.3.2 Logist

Logist software is considered one of the representative software based on item response theory (IRT) models (Linacre, 2005). Logist software introduced by Lord in 1976 brought IRT into the practical phase. Logist is widely used in the field of statistical analysis. Like other software based on IRT, it mainly deals with dichotomously scored data of (0 or 1). The parameter estimation method used by Logist is also the maximum likelihood (ML). This is because for 0-1 data variables are binomially distributed, so it is more appropriate to use the ML estimation method. Same as BILOG, Logist requires programming of command file, iteration, and achieving convergence to a pre-determined value. It is worth mentioning that Logist models have a wide range of applications and impact in several research fields. With the continuous development of computer and network technology, Logit models will also continue to play an important role in areas such as language testing and statistical analysis.

3.3.3 PARSCALE

The PARSCALE[8] software was originally developed jointly by Dr Eiji Muraki, who worked at ETS in the United States and Professor R. Darrell Bock of the University of Chicago. The software is known as the "IRT item analysis and test scoring system for scale data" (Eiji Muraki and Bock, 1997). Like BILOG, PARSCALE is also very powerful. In addition to test item equating (the command file for test equating is almost the same as that of BILOG, but the data format is different), two-valued item analysis, including one-, two-, and three-parameter logistic models, etc., PARSCALE is also capable of multi-valued item analysis, e.g., Samejima model, partial scoring model.[9] PARSCALE and BILOG are almost the same in terms of program writing and command file format. In other words, users who know how to program with BILOG command file can also know how to use the PARSCALE system easily and comfortably (Zhang, 2004). The author believes that, in general, PARSCALE is currently used more in other areas than in language testing.

3.3.4 WINSTEPS

WINSTEPS is a software program developed by Mike Linacre in the early 1980s for Rasch analysis. It focuses on measuring latent traits such as abilities or attitudes. WINSTEPS allows users to analyze data from assessments and surveys to evaluate item difficulty, person ability, and other parameters that are central to the Rasch model. The software is widely used by researchers and practitioners for its robust capabilities in scaling and scoring assessments, ensuring that measurements are valid and reliable. Over the years, WINSTEPS has been continuously updated and refined to incorporate advancements in the field and to meet the evolving needs of researchers and practitioners. The program has become a widely used tool for conducting Rasch analyses in educational and psychological measurement.

Similar to SPSS, WINSTEPS is also very powerful. It can perform all kinds of tests, model estimation, graphing, and can also conduct exploratory factor analysis, principal component analysis, and many other statistical means. In the case of the test equating, the anchor item (IAFILE=) can be set for test equating (WU and Zhang, 2016). In terms of data import, WINSTEPS is compatible with Excel, R, SAS, SPSS, and STATA data and programs, which can be easily converted to WINSTEPS instruction formats and data files. Besides, WINSTEPS has come to realize its long-time default (English) interface (must have brought confusion or embarrassment to) Chinese-speaking users in the process of upgrading and updating and has added the function of setting Chinese interface.

Another thing worth mentioning is that WINSTEPS is promoted far better than other software counterparts. In particular, the "HELP" function offers over 100 help files with examples and detailed explanations of output forms, for users' reference. The "topics" covers analyses, and Q & A. For professional users in particular, the Bond and Fox examples available on winsteps.com[10] are very helpful. In addition, for beginners without any Rasch background who encounter unfamiliar English terminology related to Rasch measurements, there is also a website at[11]; there is even a short video introduction of a few minutes as well as a course data file at[12] that provides information the user may need. Of course, this shows Mike Linacre has done a good job!

3.3.5 TAM

The TAM[13] system is Test Analysis Modules, commonly known as "TAM", developed by Alexander Robitzsch, Thomas Kiefer, and Margaret WU using the R-language. In terms of estimation methods, TAM, like BILOG and PARSCALE, performs marginal ML estimation and joint ML estimation, including unidimensional and multidimensional item response models, the Rasch model, the two- and three-parameter models of IRT, the generalized partial model, the multidimensional Rasch model, and so on. TAM system fits well with Rasch model and two-parameter model. The greatest advantage of TAM is free to download from the internet as long as R studio (also free) is installed on the personal computer. Another great feature of TAM is that TAM runs faster. Back in 2013 when PROMS2013[14] was held at Sun Yat-sen University in Kaohsiung, China, the conference organizers invited Margaret WU to run a two-day pre-conference workshop featuring the TAM system[15] (see Chapter 6, 5.5.2 for details).

3.3.6 RUMM

RUMM is considered one of the representative Rasch-based software (Linacre, 2016). RUMM stands for Rasch Unidimensional Measurement Models and was developed in VB by D. Andrich, B. Sheridan in Australia, Luo Guanzhong in China (2001). RUMM2030 Getting Started Manual provides many features for item analysis; for example, 5000 items can be analyzed and processed at one time, and up to 9 individual factors (or sample characteristics) can be processed for each item with Facet activated, and up to 9 distractor items can be processed for each item when analyzing and processing multiple-choice data. Up to 9 distractors can be processed for each item when analyzing and processing data based on multiple-choice questions. Like other similar software, the core of RUMM is conducting parallel test equating (Andtrich et al., 2001).

3.3.7 ConQuest

The ConQuest[16] developed by Mark Wilson is considered one of the representative Rasch-based software (Linacre, 2016). ConQuest is equipped with the following features: Handles well the simple logistic model of the Rasch model, the rating scale model proposed by Andrich (1978), the distributional scoring model proposed by Masters (1982), and the ordered interval model proposed by Wilson (1992). In terms of parameter estimation methods, as with other similar software, ConQuest can implement marginal ML estimation, quadrature methods, and Monte Carlo methods.

In addition to this, ConQuest performs the calculation of reliability coefficients based on CTT, item analysis of distractors, and difference in Item Function (DIF) tests. It can also analyze the degree of consistency between multiple raters, as Facet does. Like BILOG, PARSCALE, and others, the data files and command files of ConQuest are also separate. The data file is used to edit and set the format of the data, specify the name and location of variables, etc. The command file is used by the user to write commands to process the data according to the users' requirements. When using ConQuest software to import external data, such as SPSS format files or EXCEL data files, users need to change the column width of the original file data, and save the data file as a separate file name with the suffix "-.dat". In terms of data processing, ConQuest software also has the function of processing multiple data types at once. The operation is simple and convenient. ConQuest is considered as a kind of more practical software in the field of psychometrics.

3.3.8 GITEST

GITEST was developed in earlier days by the late Chinese Professor Gui Shichun and Li Wei of Guangwai in Guangdong province, China (Gui Shichun, 1986; Li Wei, 2000; Zhang Zhigan, 2021; Wu and Zhang, 2016). The software written in BASIC has two main functions: item analysis based on CTT and test equating based on the Rasch model. GITEST runs under DOS on Mac and can be used for data processing for large-scale tests, for classroom teaching demonstrations, and for solving practical problems in language testing courses. In terms of data processing, GITEST has its own data editor, which accepts data in binary scoring mode and raw data types of integers and chars. GITEST analyzes each part of a whole set of test paper, producing a detailed report for users' reference, good at handling a data matrix of 200 items X 10,000 test takers at a single GITEST run. The greatest contribution of GITEST to language testing research in China is that the software was used to successfully calibrate all the test items and conduct test equating for the "ten-year (1990–1999) MET Equating Project sponsored by the Examination Centre under the State Education Commission of China" and no error occurred at the technical level in terms of parameter calibration and ability estimation. In spite of these features, with the rapid development of computer science and technology, network technology and artificial intelligence, GITEST has been replaced by the RASCH-GZ system in China, indicating that the research and application of the Rasch model in China have fully kept up with the times.

3.3.9 Rasch-GZ

Rasch-GZ[17] is Rasch-based Item Analysis and Test Equating system/online platform successfully developed by a qualified research team consisting of computer engineers, language testing professionals, and Rasch experts organized by the author himself from 2019 to 2022 based on GITEST. It is the first Rasch-based software launched in China that provides interface and generates outputs in either English or Chinese subject to users' research background.

From a technical point of view, what characterizes Rasch-GZ can be summarized as follows: first of all, the use of network language programming Java, Python, HTML, Delphi). In terms of data processing, the greatest enhancement is the data import compatible with EXCEL, which can handle the data matrix of unlimited items by unlimited subjects, while the export file can be a WORD or PDF document, which is easy to download for thesis writing and research report, etc. The plotting function of Rasch-GZ is also good. At the same time, Rasch-GZ provides online technical support. Regarding error tolerance, the Rasch-GZ system is incorporated with the current AI functions. For example, it provides instant feedback (either in written or spoken tips) on data format entry, guiding beginners until the successful completion of data entry. In addition, the VARIMAX and other problems are solved in programming. With friendly yet simple menu interface to operate, users needn't write any command file. In the test equating part, for example, the number of linking items can be adjusted online (either increasing or decreasing), and the linking items that do not meet the requirements after Chi-square test will be automatically deleted.

Finally, the Rasch-GZ has other two satisfactory conclusions:

(1) The data algorithmic schema has once again confirmed the hypothesis first formulated by Wright and Stone (1979) in "Best Test Design", "i.e., to rephrase here: the facility index of linking items turns out to be **Difficult** in an easy test, while the ones turn out to **Easy** in a difficult test".[18]

(2) Rasch-GZ basically encompasses the solutions to practical problems that may be encountered in test-equating with the Rasch model, and

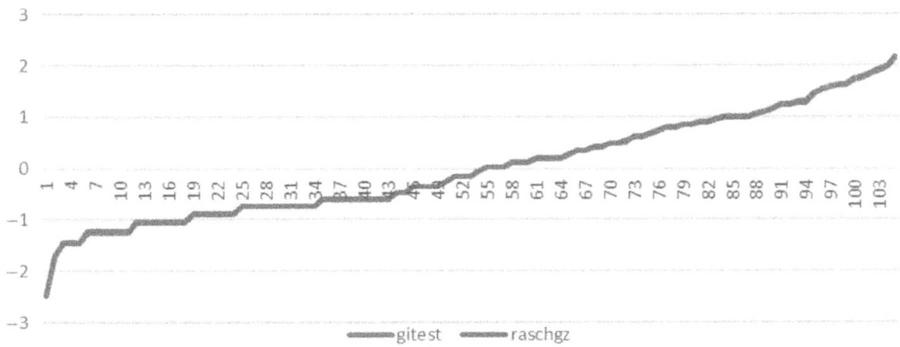

FIGURE 3.2
Two identical curves based on the same set of data produced by Rasch-GZ and GITEST analyses

the resulting estimates of item parameters, scale scores, and so on are accurate and stable. After the development of Rasch-GZ was completed, the author deliberately used both Rasch-GZ and GITEST to calibrate the test items of the same set of representative parallel tests and to analyze by comparison the difficulty parameters of the two tests. Figure 3.2 shows the two overlapped curves of the test item parameters of the same set of tests produced by both Rasch-GZ and GITEST.

The successful development of the RASCH-GZ system indicates that Chinese scholars have occupied a place in the international Rasch modelling research field, and so to speak, also represents the current status and ongoing development trend of Rasch modelling research in China.

3.3.10 R Package Dexter

The R package Dexter is actually a framework designed for psychometric analysis, particularly focused on educational and psychological testing data such as exams or assessments. There are five key features of the Dexter package:

(1) **Data Management**: Dexter provides data management functions to import and organize test data, and supports multiple psychological measurement methods, such as generalized models for multiple-choice questions, exploratory and confirmatory DIF analysis, etc.
(2) **Dichotomous Scoring**: Dexter facilitates the scoring of test responses, where responses are dichotomously scored data, i.e., yes or no; correct or incorrect type in nature, thus allowing users to calculate total scores and perform other basic scoring operations.

(3) **Model Support**: Dexter supports 1PL of IRT/Rasch and estimating ability. However, currently the package does not support 2PL and 3PL models (see 1.9).
(4) **Item Analysis**: The package generates detailed analyses of test items and overall test performance, helping users in understanding item difficulty, discrimination, and other psychometric properties (see 2.1).
(5) **Item Moderating:** All the reports and visualizations to summarize the analysis results generated by Dexter offer efficient guidelines for item moderating (see 2.1).

There are many ways to install and to use Dexter, for example, the most common one is to install it directly through CRAN. Install the Dexter package using the following code:

```
RCopy Code
install.packages("dexter")
```

3.3.11 myRasch

myRasch is a software tool designed for Rasch analysis for assessments and questionnaires. Developed by Dr. Ahmad Zamri Khairani, like other Rasch-based software, some common features[19] of myRasch include:

(1) **Item and Person Analysis:** myRasch provides detailed analysis of both test items and test takers, helping to identify how well test items discriminate between different levels of ability and how test takers perform relative to the scale.
(2) **Fit Statistics:** myRasch offers infit and outfit statistics for both items and persons to assess how well the data fits the Rasch model.
(3) **Graphical Output:** myRasch generates graphical representations of data, such as item characteristic curves, person-item maps, and other visual tools to provide the most intuitive data analysis curves.
(4) **Parameter Estimation:** myRasch calibrates both item difficulty and person ability parameters, providing insights into the relative difficulty of test items and the abilities of test takers.
(5) **Differential Item Functioning (DIF):** myRasch ensures fairness in assessments by specifying items that function differently across different groups.
(6) **Scalability:** myRasch supports the analysis of both small and large data metrics, making it suitable for a wide range of research and practical use.
(7) **User-Friendly Interface:** myRasch simplifies the process of conducting Rasch analysis to be accessible to users with varying levels of statistical expertise.

3.3.12 BEAR

BEAR stands for "Berkeley Evaluation and Assessment Research". The system was developed by Mark Wilson at the University of California, Berkeley in the 1990s. The BEAR Assessment System was actually part of an effort made by the developer to create more effective and meaningful assessment practices in education by integrating research on learning and assessment.

Designed to address the limitations of traditional assessment methods, the system provides a more comprehensive understanding of student learning through the use of learning progressions and construct maps. The system includes the following five components:

(1) **Learning Progressions:** These refer to sequences of skills or concepts that students typically follow as they develop understanding in a subject area. These concepts help educators identify where a student is on their learning journey and what they need to learn next.

(2) **Assessment Tools:** These refer to the use of assessment tools that are aligned with learning progressions. Such tools are designed to provide detailed information about student learning and to support instructional decision-making.

(3) **Construct Maps:** These refer to visual representations outlining the skills and knowledge students are expected to develop. These skills serve as guidelines for both teaching and assessment.

(4) **Item Design:** The system advocates for the creation of assessment items that are closely aligned with the construct maps and learning progressions, thus guaranteeing assessments accurately measure student understanding and progress.

(5) **Outcome Space:** This component defines the range of possible student responses and interprets these responses to know how well student learning is going.

3.3.13 Lexile Framework

The Lexile Framework[20] for Reading is a system developed by Prof. Jack Stenner of MetaMetrics, an educational measurement and research organization in the late 1980s. The Lexile was introduced to the public in the early 1990s. It has since become a popular tool for educators and parents to assess and support students' reading progress and also used to measure both a reader's ability and the complexity of a text. The Lexile provides a standardized metric, known as a Lexile measure, which helps match readers with texts that are at an appropriate difficulty level for their reading skills. The Lexile Framework is widely used in educational testing settings to support reading development and comprehension and good for designing reading comprehension test. According Jack Stenner (2013), the designer, The Lexile Framework includes the following four features:

(1) **Making Test Scores Actionable**
Today's students take many different types of assessments and receive many different scores. These scores provide important measures of student proficiency in content areas, but they are often static. The data can't be used to inform instructor or to help educators and parents select appropriate reading materials based on each child's ability. The Lexile Framework for Reading changes that. When an assessment is linked with the Lexile Framework, students' test scores immediately become actionable. A Lexile measure is the most widely adopted reading metric, measuring both reader ability and text difficulty on the same scale. Lexile measures are powerful tools for linking assessment with instruction across the curriculum, at home, and in the library, by taking the guesswork out of selecting materials that can help to improve student reading ability.

(2) **Matching Readers with Text**
According to the designer, if you need to buy a pair of basketball shoes for your kid, you do not buy shoes by age, you buy them by size. However, for years, we have matched students with books and other learning materials based on age or grade level. If a student likes science fiction books and is nine years old or in fourth grade, he is given "fourth-grade science fiction" to read. But what if that fourth grader's reading ability is far higher than the "average" student his age, or he is not quite reading on a fourth-grade level? Like your kid's age-12 basketball shoes, the text simply does not fit the student. In this sense, the Lexile Framework determines a student's reading ability, not his grade level. With Lexile measures, assessment results are used to ensure a "good fit" because the measure is used to select reading material that meets and challenges each student's ability.

(3) **Measuring Student Growth on a Common Scale**
Many reading comprehension assessments and programs are linked with the Lexile Framework. This offers educators and parents a common scale for monitoring progress throughout the school year and their entire education, regardless of what assessments they take. A Lexile measure can provide the same continuity for reading growth that families have when they mark a child's height on the wall with a pencil. Like the wall, the scale never changes, so progress is easy to see even if the student changes grades or takes a different assessment. Lexile measures range from below 200L for emergent readers and text to above 1600L for advanced readers and text.

(4) **Connecting Students with Lexile Measures**
Lexile measures may already be available for your students. All major US standardized reading tests and many popular international tests

have been linked with the Lexile scale including ETS's TOEFL iBT test, TOEFL Junior test, and TOEIC Listening Reading test. Lexile measures are also reported from the Scholastic Reading Inventory. Recognized as the standard for matching readers with texts, tens of millions of students worldwide receive a Lexile measure that helps them find targeted readings from more than 100 million articles, books, and websites that have been measured.

(5) **More Information, Not More Testing**
Best of all, the Lexile Framework adds value to state or classroom assessments—adding more information, not more testing time. It is not another test or a reading intervention program. Lexile measures provide a thermometer for measuring students' reading abilities that talented educators, involved parents, and motivated students using to improve learning. Lexile measures tie day-to-day work in the classroom to critical high-stakes tests. The Lexile Framework offers a "big picture" view of growth of student reading ability from pre-school through graduate school.

3.4 Comments on the Computer Software Programs

Based on the accumulated experience in terms of using both IRT- and Rasch-Based software over the past years, the author of this book should say technically, all the software listed in (Section 3.7) are powerful, stable in performance, and accurate in estimation. The author once used BILOG, PARSCALE, and GITEST to calibrate the difficulty of the same set of MET data during the implementation of the ten-year (1990–1999) MET Equating Project sponsored by the Examination Centre under the Ministry of Education, China and the resulting parameter curves are shown in Figure 3.3 (Quan Zhang, 2004).

As seen in Figure 3.3, the three curves generated by BILOG, PARSCALE, and GITEST, respectively, are going very close, and the two ones by BILOG and PARSCALE are almost overlapping. Our interpretation goes thus: This is related to the different iterative convergence values set in their respective command files; BILOG came to convergence at (CRIt = 0.005) after 6 iterations, and PARSCALE converges at (CRIt = 0.01) after 72 iterations. GITEST is slightly different from the two. This is because all the default values of GITEST are fixed in the program. It can be seen that BILOG, PARSCALE, and GITEST are stable in terms of parameter estimation, with a correlation of (r = 0.93), which is the reason why we just used GITEST since then.

From the author's point of view, in the current practice of using software for test equating, regardless of BILOG, PARSCALE, WINSTEPS, or RUMM, there exists a common feature, that is, the command file and the data file are separate. To be more exact, the users are supposed to not only edit the data

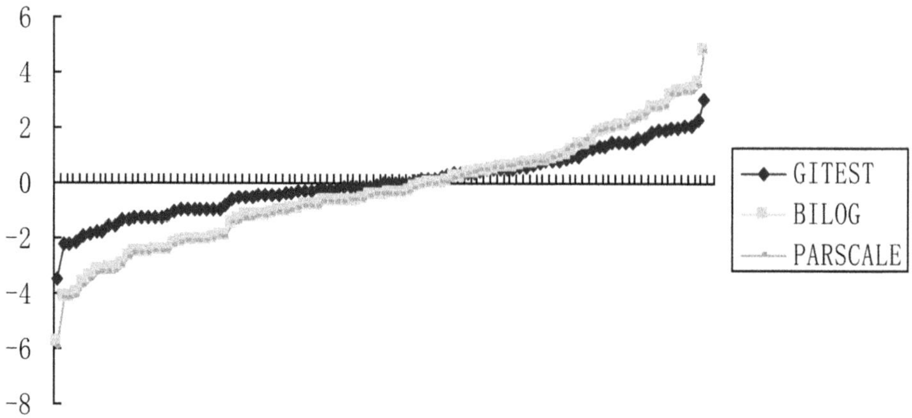

FIGURE 3.3
Curves of test items generated by GITST, BILOG, and PARSCALE using the same set of MET data (1990–1999)

file according to the required data format, but also write the command file according to the syntax of the command file format. In fact, it is writing a short computer program (see Appendix I and II provided at the end of this chapter) because the computer system reads the commands and runs to do the tasks according to the command file. At this point, software designers may not have been aware that even such a short format programming would be difficult for users who are lack of any experience of computer programming (especially Chinese teachers of English, etc.). Another underlying reason is that while programming, technical problems may occur, such as programming errors, program abort, etc., these problems are often not resolved in a timely manner. Although BILOG gives an error code, and PARSCALE has a trial run that generates a Phase 0 output file, the results are not effective in practice. This is the very reason why we have only a small band of highly professional users confined within the language testing and psychometric community in China. They are very proud of it because they know very well the promotion of such a kind of software goes difficult among other counterparts with Chinese traditional mindset.

3.5 Translation Works

In addition to the application of test equating techniques, Rasch experts are also aware of the overwhelming need to promote Rasch measurement in Mainland China is doing translation work. In order to have this task accomplished, two pillars are necessary. One is a qualified team of translation who must possess a high level of understanding of Rasch measurement and bilingual proficiency so as to ensure the academic merits of the book could be accurately delivered from English into Chinese; the other is an appropriate Chinese publisher with a deliberate academic taste and prestige that could endorse the value of the book in the eyes of its Chinese readers (Bond, 2016). This section mainly introduces significant contributions made in this line. These translated works not only greatly promote and disseminate the Rasch measurement theory, but also, so to speak, fill the gap in today's Rasch measurement research.

3.5.1 JAM Book Abstract Translation

3.5.1.1 Constructing Variables. Book of Abstracts Volume (I) and (II)

The Rasch model is known to be the most versatile, powerful and practical. However, descriptions of such versatile, powerful, and practical approaches are mostly written in the English language. The understanding, dissemination, and flexibility of the model are never fully experienced by a majority of Chinese-speaking researchers. Such a situation existed in the academic circle for quite a long time. Motivated by this and led by Prof. Magdalena Mo Ching MOK from Hong Kong Institute of Education (now as the EdUHK), the editors first met in 2013 to discuss the possibility of publishing Rasch translation works, with permission by Dr. Richard M. Smith, into both simplified and traditional Chinese, the English abstracts related to Rasch-based research from the *Journal of Applied Measurement* (JAM), a scholarly peer-reviewed journal of measurement applications with an emphasis on the family of Rasch measurement models. The goal was to produce books that would give readers in Chinese-speaking research communities general information regarding Rasch measurement conducted in other parts of the world.

The Constructing Variables and JAM Book of Abstracts (Vols. I and II) are translated from English into both classic and simplified Chinese and published in 2014 and 2015, respectively (Mok and Zhang, 2014, 2015). These two volumes contain 410 abstracts from JAM germane to Rasch-based research work translated from English into both Simple and Classic Chinese by 45 highly competent translators who are working in 13 different organizations, universities, or institutes located, respectively, in China Mainland, China Hong Kong, China Macau, and China Taiwan.

Each of these abstracts deals with Rasch measures in their research field, covering a variety of issues ranging from education, psychology, management, testing to medicine. The abstracts also serve as valuable resources for researchers and students of non-English majors in Mainland China and Chinese-speaking researchers to be able to conduct their Rasch analyses as well as understand and critically review Rasch-based research publications (Zhang, 2019).

3.5.1.2 JAM Book of Full Paper Translation

Even though the abstract books are published, the author/editors realize that the versatile, powerful, and practical approaches of Rasch are still far from being widely understood or used outside a relatively small group of Raschers. Full paper translation is needed to meet such a situation in China. From the beginning of 2017, the team prepared to systematic translation of the representative papers from the Understanding Rasch Measurement Series in the JAM. The translators put their relentless efforts into making sure that the original concepts, issues, and examples presented to help explain various principles and applications in Rasch are reflected in the translated versions. In 2018, at Technology-Enhanced Assessment Conference in Hong Kong was launched a ceremony for a new book of full paper translation entitled Introduction to Rasch Measurement in both traditional Chinese and Simplified Chinese versions. The original authors of the book in English are Everett V. Smith Jr[21] and Richard M. Smith.[22]

The full paper translation contains 24 chapters deliberately selected from the previously published *Journal of Applied Measurement*, under the Understanding Rasch Measurement series and addresses three major areas: Theory, Models, and Applications. All the authors are experienced Raschers. Dr. Richard M. Smith stated in the Theory Section, containing the basics of Rasch measurement and how Rasch measurement differs from IRT. The chapter by Wright and Mok presents the Rasch models commonly used in the social and health sciences.

The two chapters by Linacre discuss estimation methods for these models. Estimation of model parameters is logically followed by a discussion of the assessment of the fit data to a model in a chapter by R. Smith. E. Smith then presents an overview of how the traditional conceptualizations of reliability and validity can be addressed in the context of Rasch measurement. Finally, the chapters by Wilson and Andrich discuss the underlying philosophy and rationale of Rasch measurement and why some researchers prefer to use Rasch models rather than IRT models. In reality, each chapter is very informative, and some, if expanded, would be important topics not only for newcomers but also for postgraduate students (Zhang, 2019).

3.5.2 Monograph Translation

3.5.2.1 Applying The Rasch Model by Trevor Bond and Fox (2nd Ed.) (B&F2 Thereafter)

According to Bond and Zi Yan (2016), in 2005, three influential professors in China Hong Kong have undertaken to coordinate the publication of B&F2 in Mandarin Chinese, but due to hindrances to publication, they were unable to achieve their goal. This is the first attempt to translate Rasch's original work in China. B&F2, cited more than 1,900 times, facilitates a deep understanding of the Rasch model. In particular, the authors present an accessible overview that does not require a mathematical background. This would be more beneficial for researchers of social sciences and humanities. Highlights of B&F2 can be briefly summarized in the following.

The authors present more learning tools to strengthen readers' understanding. To allow readers to select the level most appropriate for their needs, the authors divided chapters into basic and extended understanding sections, thereby providing more in-depth investigations of key topics. Apart from this, the authors emphasized the role of Rasch measurement as a priori in the construction of scales and its use post hoc to reveal the extent to which interval scale measurement is instantiated in existing data sets. Importance has also been attached to interval-level measurement data and to the demonstration regarding how Rasch measurement is used to examine measurement invariance (Bond and Fox, 2015)

3.5.2.2 Constructing Measures: An Item Response Modelling Approach by Mark Wilson

This book is one of Mark Wilson's representative works, translated by a research team headed by Professor Huang Xiaoting at Peking University, China and published by Hunan Education Press, China. The translated works (Huang Xiaoting, 2019) are the core achievement of major projects funded by the National Social Science Funding of China. The original author has served as the chairman of the American Council for Educational Measurement and the International Society for Psychometric Measurement, a member of the National Academy of Education in the United States, and was awarded the Lynn Prize by the American Educational Research Association and the Lifetime Achievement Award presented by the California Educational Research Association. The author participated in the founding of the journal Measurement: Interdisciplinary Research and Perspectives, and is also an editorial board member for multiple journals, such as *KEDI, Journal of Educational Policy, Australian Journal of Education, Psychometrika, JAM,* etc. Mark is also an authentic senior scholar in the Rasch family.

The author initially wrote the book because he taught students a course called ED274A on "Measurement in Education and Social Sciences" at the University of California, Berkeley (UCB). In designing the course, the author decided to adopt a new approach. According to him, in general social science majors, introductory courses in measurement mainly introduce students to CTT and how to use mathematical formulas to calculate reliability and validity. Although such courses often involve the construction of tests, there is often no systematic explanation on how to design scales. The ED274A course attempts to make up for this deficiency. In this course, the author attempted to teach students a method for designing scales and verifying their reliability and validity, namely, the BEAR assessment system (BAS; Wilson and Sloane, 2000), and enable students to better understand relevant measurement theories in the practical process. In order to achieve the integration of theory and practice, the author provides a new method for beginners, expecting students will be familiar with the concepts and processes related to measuring tool design and evaluation, thus laying a solid theoretical foundation for further exploration of measurement theory and technology.

3.5.2.3 A Course in Rasch Measurement Theory by David Andrich and Ida Marais

The translation commissioned by Prof. Luo Guanzhong is currently in progress. The authors apply Rasch measurement theory to the fields of education, psychology, sociology, marketing, and health outcomes in order to measure various social constructs. They focus on the first principles of both the theory and its applications. The software used in the book which is helpful in working through the text is RUMM2030 (Section 3.3.6).

To achieve the goals, the authors equip researchers with the confidence they need in order to be in control of the analysis and interpretation of data, and to make professional rather than primarily statistical decisions mechanically. The book is based on courses that have been taught both online and in intensive form for over two decades. Much of the content is based on research conducted by the two authors and their colleagues (David Andrich and Ida Marais, 2019).

Apart from serving as good resources for Rasch measurement research, the translated books can also be used as a textbook for postgraduate students of applied linguistics in general (Zhang and Zhang, 2015).

3.6 Computer Software Development

To meet the on-going rapid development with an emphasis on further promoting the use of the Rasch measurement in China so as to be well geared with the international practice today, led by Prof. Quan Zhang, a group of Chinese professionals met during the COVID-19 pandemic period since 2019 and successfully developed RASCH-GZ (3.3.9.), the first Rasch-based item analysis and test equating system which produces output files in both Chinese and English. In this way, the understanding, dissemination, and the beauty of Rasch model, or to be more exact, the Rasch-based research has been fully experienced or shared by a majority of other Chinese-speaking researchers, especially those young researchers whose majors are not in English. The features of Rasch-GZ can be briefly listed below:

(1) Capable of processing unlimited dichotomous data matrix;
(2) Adjustable number of linking items prior to test equating;
(3) Chi-square test prior to test equating;

(4) Plotting and online technical supports;
(5) Language (Chinese or English) can be selected;
(6) Incorporated with AI, both voice and text hints are provided to guide the users.

The significance of developing Rasch-GZ lies in the following four aspects and is highly in conformity with the PROMS goals (Section 6.1.4.):

(1) Rasch-GZ is the first Rasch-based system for item analysis and test equating which generates output files in both Chinese and English;
(2) The Rasch-GZ interface designed has well met the needs of Chinese-speaking researchers;
(3) The release of Rasch-GZ greatly facilitates the learning and application of Rasch model among Chinese learners;
(4) Student version is free of charge, encouraging younger researchers to use Rasch model to conduct research.

3.7 Research Background and Use Cases of Rasch in Malaysia

According to Zali (2024),[23] the research and development of Rasch in Malaysia fall into three aspects: teaching, workshops, and research.

3.7.1 Rasch Teaching

The earlier pioneers who were spreading the Rasch measurement model to academic communities throughout Malaysia are Abdul Halim Ahmad and two other Education Ministry officers who visited Ben Wright in Chicago in 1980. But it was Prof. Wan Rani Abdullah who started to popularize Rasch in the academic community in Malaysia. Later, other practitioners started teaching the Rasch model. From 1996 on, Prof. Dr. Vincent Pang and Yap Yee Khiong conducted courses on Rasch Model for Research using BIGSTEPS at the Education Technology Division, Sabah State Education Department. The teachers of Rasch who disseminate knowledge on Rasch measurement model such as Prof. Rosseni Din (2009), Dr. Akbariah (2008), Prof. Noorlide Abu Kasim (2007), Saidfudin Masodi, Dr. Azrilah, and Prof. Azami Zaharim (2007) are synonymous to Rasch teaching in Malaysia.

In 2010, myRasch (Section 3.3.11) was first introduced in Malaysia. Developed by Dr. Ahmad Zamri Khairani, a Malaysian academic and expert in educational measurement, myRasch created the online educational platform to facilitate the teaching and application of the Rasch model in educational assessments. Dr. Khairani's work on myRasch aimed to provide educators with a practical tool for analyzing student performance data, supporting data-driven decision-making in education, thereby improving teaching strategies. It is believed that the platform is particularly useful in identifying areas where students may need additional support or where instructional methods can be enhanced. By leveraging the Rasch model, myRasch aims to provide a more accurate and objective measurement of student abilities compared to traditional testing methods.

3.7.2 Rasch Workshops

Apart from teaching, Malaysian Psychometric Association (MPA), myRasch, and individual trainers had conducted Rasch measurement workshops since early 2005. Some universities began to run workshops of Rasch measurement model as well. During the early 1990s, Prof. Dr. Vincent Pang from Universiti Malaysia Sabah started using QUEST and BIGSTEPS to conduct item analysis using Rasch and conducted Rasch analysis workshop together with Nuinda Alias, Mohd Said Hasran, and Hashim Asman in Gaya Teacher Education College in Kota Kinabalu. Universiti Teknologi MARA (UiTM) and Universiti Industri Selangor (UNISEL) followed the suit in 2012.

3.7.3 Rasch Research

It is generally believed that Prof. Wan Rani Abdullah PhD had made some impact on studies using Rasch in Malaysia. Zali (as cited in Ben Wright, 1989) in the Notes on the 1989 MESA dissertations:

> In educational administration, Wan-Rani Abdullah, Senior Research Officer, Educational Planning and Research, Malaysian Ministry of Education, developed objective measures of Malaysian teaching styles, the definition of which clarified the working hierarchy of teacher responsibilities and brought out significant differences between urban and rural teachers. Now Wan Rani is applying "the knowledge and ideas which I gathered at U. of C., especially Rasch measurement and HLM" [hierarchical linear modelling], to his UNESCO project on educational wastage and his study of performance differences in languages, mathematics and science.

Prof. Dr. Wan Rani's earlier contribution to Rasch measurement was also noted in his paper (1989) entitled One Is Sufficient or Two Are Necessary, a substitute for "Clarifying Variables of Teaching through Factor Analysis of Rasch Standardized Residuals". Prof. Wan Rani later moved to Universiti Sains Malaysia where he introduced Rasch measurements as a proven methodology in research studies. In the later 1990s, two other notable Rasch practitioners are Khoo Sek Teun and Dr. Haniza Yon, the latter studied Rasch in Michigan State University in 1998, under Dr. Edward Wolf, a student of Prof. Mark Wilson. Currently, Dr. Haniza is the president of MPA. However, only after 2005 does the number of research studies in Malaysia increase.

From 2010 onward, many Rasch-based MA and PhD students began to graduate. According to Zali (2024), most of their theses or dissertations were examined by examiners who were not conversant in Rasch measurement model. As such, two issues arose: One was that the parts regarding Rasch analyses were not checked professionally in details for correctness; the other was that examiners requested their students to remove from their papers the inappropriate citations of the Rasch analyses. To support teaching and research in this line, several books on Rasch in Bahasa Malaysia (i.e., Malay language) were published. Among them are books on Basic Rasch Measurement Model[24] by Dr. Azrilah, Saidfudin, and Prof. Azami Zaharim, and Rasch Model Application[25] by Dr. Bambang Sumintono. Apart from this, two main academic organizations supporting Rasch study and teaching were established in 2017. They are MPA headed by Dr. Haniza Yon, and myRasch led by Prof. Nazlinda Abdullah, maintaining the Facebook forum—Rasch Measurement Model Malaysia Chapter. However, to use Zali's words, more efforts are needed to spread awareness and promote Rasch measurement to be included in research methodology courses in majority of academic institutions in Malaysia and beyond.

3.8 Rasch Measurement in Other Countries and Beyond the Pacific Rim

3.8.1 Rasch Measurement Special Interest Group (SIG)

While talking about PROMS,[26] the author cannot ignore the most important academic group for Rasch measurement research and development on the other side of the ocean, which is Rasch Measurement Special Interest Group (SIG) under the American Educational Research Association (AERA). Founded in 1992 by Benjamin D. Wright in person and with 67 key members (2024), this Rasch SIG provides a dedicated platform for individuals interested in Rasch measurement models to share research, discuss developments, and promote the use of these models in research studies of various fields. A prominent organization dedicated to advancing educational research and its practical application. The Rasch Measurement SIG focuses on the use and development of Rasch measurement models, a family of probabilistic models used for analyzing data from assessments to measure variables such as abilities, attitudes, and personality traits. While playing a significant role in establishing the SIG as a platform for researchers and practitioners interested in this area, Wright was instrumental in promoting the use of Rasch measurement in educational and psychological contexts. Like PROMS, the AERA holds its annual meeting once a year. The AERA is a true international event where researchers, educators, and practitioners from around the world rub shoulders to share and discuss the latest developments and research in the field of education. Every year AERA selects a different city across the United States to host the event, providing attendees the opportunity to experience diverse settings and venues. The choice of location is announced well in advance of the meeting to give participants enough time to make necessary travel arrangements. Past meetings were held in San Francisco, New York, Washington D.C., and Chicago, among others. As a platform for researchers, practitioners, and educators interested in the theoretical and practical aspects of Rasch measurement, the Rasch SIG facilitates the exchange of ideas, research findings, and best practices among its members. Apart from this, the Rasch SIG organizes activities such as sessions and presentations at the AERA annual meetings, newsletters or bulletins publications, collaboration among professionals in the field, business meetings, and awards of various kinds such as the bi-annual Benjamin Drake Wright Senior Scholar Award[27] to Dr. Trevor Bond from Australia in 2024. The aims of the Rasch SIG are to promote a deeper understanding of Rasch measurement and their applications globally in educational and psychological research fields.

3.8.2 Rasch in Mongolia

With the continuous deepening and promotion of research and application of Rasch measurement, countries outside the Pacific Rim are also gradually showing interest.

In 2011, the 5th National TEFL/1st Mongolia TESOL International Conference held under the theme "Learning through Practice" at Ulaanbaatar, Mongolia, the author was invited to deliver the keynote speech plus a workshop. The presentation falls into three parts. First, a brief introduction was given to GITEST, BILOG, and PARSCALE, representing, respectively, Rasch-based and IRT-based software handling dichotomous data and polytomous data. The author first revealed such a phenomenon in teaching and testing: Greater importance is attached to test scores only; for example, once a test is administered and the relevant test scores are used, the test paper will be put aside, and no further analyses are taken. The author emphasized, in fact, test item analysis is more important. It is in this sense that the presenter attempted to encourage the whole TEFL field to attach equal importance to both test scores and test item analysis for improvement of both teaching and the quality of test papers. This creates chances for participants to ask questions for discussion. Next, the presenter showed how to interpret the feedback to language teaching and improvement of test quality based on simulated data. Here, the participants were exposed to examples concerning item analysis, test scoring, and equating. In particular, the presenter further emphasized that only the test scores obtained from the parallel tests, which are equated can be comparable on the same scale and thus fair to all the different test takers taking the parallel tests in different time and only the application of Rasch model or IRT can solve the problem of test equating. Finally, the presenter demonstrated with simulated data showing how each computer software runs. This makes participants immediately aware that running such Rasch-based or IRT-based computer software is NOT complicated.

In 2017, the author's paper was accepted by The First International Conference on Educational Measurement, Evaluation and Assessment (ICEMEA)[28] held by Emirates College for Advanced Education (ECAE) in Abu Dhabi, UAE. In the presentation session, the author gave a detailed introduction to Rasch model: Retrospect and Prospect in China (Zhang, 2019).

3.8.3 Rasch in Abu Dhabi, UAE

In 2007, The International Conference on Educational Measurement, Evaluation and Assessment (ICEMEA) was held by Emirate College for Advanced Education in Abu Dhabi, UAE. Prof. George Engelhard Jr. (the current chairman of PRMS) was invited as the distinguished speaker from College of Education, The University of Georgia, USA to deliver the speech: How can invariant measurement based on Rasch models inform educational assessments?

In his speech, George addressed that measurement serves numerous purposes in education. Test scores can be used for diagnostic, formative, summative, and strategic assessments. For each of these purposes, the intended and appropriate use of test scores should be evaluated in terms of their validity, reliability, and fairness. The validity, reliability, and fairness of test scores can be defined based on the requirements of invariant measurement. Invariant measurement addressed by George refers to assessments that yield test scores that "stay put while the user's back is turned". Rasch models offer an approach for developing invariant measures that can support the intended uses of educational assessments. This presentation provides an overview of these concepts, as well as the implications for research, theory, and practice related to educational assessments.

ICEMEA under the Patronage of Her Excellency Jameela Bint Salem Mesbeh Al Muhairi, Cabinet Member and Minister of State for Public Education Strategic Partner, Ministry of Education, UAE is an important conference in UAE. The organizers of ICEMEA had a very clear and super insight approach. They believe that educational development in a country largely depends on how well progress is measured, evaluated, and assessed. An effective education system needs robust quality control and monitoring mechanisms in place to efficiently evaluate the performance of the learning organizations. Considerable attention has been attached to the roles of educational measurement, evaluation, and assessment with a view to improving the education systems in general and to adequately prepare the young generation to meet the ever-growing demands of the 21st century in particular.

The conference is aimed at those working in all sectors of education and training who have a keen interest in advancing their knowledge and skills in educational measurement, evaluation, and assessment. This event will be of great interest to educational leaders, policy makers, educators, and researchers, and a key international gathering of acclaimed scholars with immense opportunities for dialogue, exchange of views, and networking in a productive environment as well. There are approximately 100 participants, including the author of the book attending the conference. They are from Australia, Canada, China, France, Singapore, Turkey, UAE, UK, and USA.

3.8.4 IACAT

IACAT stands for the International Association for Computerized Adaptive Testing. Founded by David J. Weiss in the United States in 2009, a notable figure in the field of psychometrics and CAT, Weiss has contributed significantly to the development and promotion of adaptive testing methodologies and has been instrumental in establishing IACAT as a platform for advancing research and practice in this area and for collaboration among researchers and practitioners in the field. It is known to all in the language testing field that CAT adapts to the test takers' ability level by selecting test items based on their previous responses, making the testing process more efficient and tailored to the test takers. Such application of CAT must be supported by an item bank with a large number of well-calibrated test items via test equating. All these are actually based on the Rasch model measurement theory or IRT theory. In fact, we can also consider CAT as a specific application or manifestation of the Rasch model. Like PROMS, IACAT also organizes conferences and publishes research to support the dissemination of CAT-related information. As of the latest available information in 2023, the current president of IACAT is John R. Lewis.

3.9 Summary

This chapter introduces the research background and Rasch-based use cases with focus on test equating, practice, and problem solving in China's MET, CET, and HSK, some ideas of which are disclosed for the first time in English. A more comprehensive introduction to the application of Rasch model to test equating is addressed, exploring the concepts of test equating, the conditions that should be met for conducting test equating, the concrete methods of test equating, and all the popular software used for test equating. The author argues that in today's era of big data and computer technology, true testing should be based on real and objective data rather than relying on classroom instruction or theoretical discussions. Although modern testing theories such as IRT or the Rasch model have been accepted by Chinese professionals of language testing, they are still mainly confined within seminars, conferences, or classrooms. The situation in terms of the application is by no means optimistic.

As far as the application of computer technology in language testing is concerned, the current situation in China is fully in line with international practice;

however, in all types of institutes across the country, testing theories and applications remain at the level of CTT, and are basically limited to linear or descriptive statistical analyses. In this sense, Rasch model measurement needs to be further disseminated and promoted. Therefore, the author would like to emphasize that what we are doing does not have to be more accurate, but has to be more correct! So long as the method is correct, using the Rasch model to process real data in the scientific research to draw reliable conclusions is crucial!

Specifically, today in language testing research and practice, test equating remains a prerequisite for computerized adaptive testing (CAT), item banking, and web-based or online testing. Test equating makes feasible the observation of changes in item difficulty, thus ensuring reliable measurement of the corresponding ability estimates of test takers at different times and on different occasions. The most important intrinsic meaning of test equating is to maintain test security and test score fairness. Although test developers and proposers try to construct test papers that are as similar as possible in content and form, these papers always differ in difficulty, so comparability of test scores (expressed as ability estimates) obtained from different tests is always a matter of concern. In China, the MET and CET are typical examples. In this sense, the Rasch-GZ system introduced in this book only serves as a guide to the effect of jade. We need more academic exchanges and promotions, more international exchanges and co-operation to make the Rasch-GZ system perfect. Further research in this area will be discussed in a separate article. Apart from the test equating, totally listed are thirteen kinds of computer software programs currently used in the psychometrics, the Rasch-based research fields, and particularly with comments to guide readers to use the programs properly. Important translation works are introduced. Research background and use cases of Rasch model in Malaysia and other countries and regions are also presented with some details. In particular, details are given to Rasch Measurement SIG and IACAT.

In summary, the Rasch model can be a good solution to problems of objective measurement encountered in language testing as well as in social science disciplines, and is suitable for researchers in specialized areas such as language testing. The author concludes this chapter by reminding the reader of the following two additional points:

(1) The easiest way to learn how to conduct the Rasch model measurement is to download the student version of the Rasch-GZ system and the short video user manual from the following website http://www.rasch-gz.com

The student version comes with a small data matrix of (30 items by 40 subjects) with short videos in both Chinese and English. Just follow the user guide to work at the click of a mouse. This can be helpful for Rasch modelling or classroom teaching. To conduct test equating of large-scale test with high stakes, users can request the professional version of the Rasch GZ system online.

(2) Meanwhile, it is also necessary to point out that nowadays although the testing software listed in (3.7) can be obtained either online or offline, keyboard operation, mouse clicks would make the calibration or algorithm process automatic and the results are thus obtained with ease; however, in a strict sense, this is only the aids for test equating because the users could not observe the actual process of calibrations. What is obtained is the product: i.e., the specific parameters. Sometimes, we may not interpret the results very well, or sometimes, our understanding of the parameters and the interpretations may not be correct. This means that we still need to be serious about the estimation process and that we still need to carefully read some professional books to master the basic concepts and principles. This is why the author has listed the relevant authoritative books at the end of each chapter for readers' reference. There is a Chinese proverb which goes: Over time, the results will be seen. Or to say in French, pétit à pétit, l'oiseau fait son nid.

Appendix I

BILOG instruction file for parallel test equivalents[29]

```
> GLOBAL NWGHT=3, DFN= 'MET-95.dat', NPA=3;
> LEN NIT = 50.
> INPUT FOR = 2, NTOt=30, NALt=4, INOpt=2,NIDc=11
  (11A1, 11, F2.0X, 30A1)
> FOR LEN=50, ITE=(1(1)50).
> FOR LEN=50, ITE=(1(1)10,31(1)50).
> TEST TNA=MET95LNK; LINK = (1,1,1,1,1,1,1,1,1,1,1,1,0,0,0,0,
  0,0,0,0,0,0,0,0,0,0,0,0,0,0,0,0).
> CALi SORt.
> SCOre MET=1,RSC=2.
```

Appendix II

PARSCALE command file for estimating the DIF model[30]

```
> COMMENT.
> FILE DFN='PSLDAT\EXAMPLE06.DAT', NFN='PSLDAT\EXAMPLE06.NFL';
> INPUT NIDW=8, NTOT=6, LENGTH=6, NTEST=1, NFMT=1, MGROUP=2;
  (8A1, 5X, A1,/,6A1)
> TEST TNAME=PARV3E, ITEM=(1,2,3,4,5,6)
  INAME= ('I001','I002','I003','I004','I005','I006')
  NBLOCK=6;
```

```
> BLOCK1 REPEAT=6, NIT=1, NCAT=3, ORIGINAL= (1,2,3);
> MGROUP DIF=(0,1,0,0), GNAME=(MALE, FEMALE), GCODE= ('1', '2');
> CAL LOGISTIC, PARTIAL, NQPT=25, CYCLES= (100, 1, 1,1,1, 1),
  NEWTON=2,
  CRIT=0.01, DIAG=0, SCALE=1.7, POSTERIOR.
> SCORE
```

Notes

1. For more information on the solution and the specific implementation of the 10-year MET Equating Project, see Shichun Gui et al. (1993).
2. http://www.CNKI.net
3. Due to space and big data, the authors will publish a separate paper on the specific approach to the ten-year MET Equating.
4. There is also a number of methods for test equating for which the interested reader is referred to Samuel A. Livingston (2004). The Chinese practice of MET equating is very early (1990–1999) so there were no many reference materials for it.
5. See Samuel A. Livingston (2004). Equating Test Scores (Without IRT). (pp. 17–35) ETS. Princeton, USA
6. For a detailed discussion about test equating in large-scale examinations both at home and abroad, the reader is referred to the bibliography at the end of this chapter.
7. PC-BILOG: Item Analysis and Test Scoring with Binary Logistic Models.
8. PARSCALE: IRT Item Analysis and Test Scoring for Rating Scale Data is now owned by Scientific Software International (SSI). The official website is http:// www. ssicentral. com/irt/.
9. Namely: Graded Response Model and Partial Credit Model.
10. http://www.winsteps.com/forum
11. http://www.winsteps.com/winman/glossary.htm
12. http //www winsteps com/a/course-data zip
13. Test Analysis Modules (TAM). For more information: https://doi.org/10.32614/CRAN.package.TAM
14. See: http://www.education.nsysu.edu.tw/TERA-PROMS2013/
15. For a detailed description of TAM and R-Studio, the interested reader is referred to the following:

    Adams, Wilson and Wang (1997). https://doi.org/10.1177/0146621697211001;

    Adams, Wilson and Wu (1997). https://doi.org/10.3102/10769986022001047;

    Formann (1982). https://doi.org/10.1002/bimj.4710240209;

    Formann (1992). https://doi.org/10.1080/01621459.1992.10475229.

16. Readers are welcome to visit https://www.acer.edu.au/conquest
17. For details, readers are welcome to visit : https://www.rasch-gz.com
18. See Wright, B.D., & Stone, M.H. (1979). Best Test Design Rasch Measurement. MESA. (p. 90) Originally published as: The linking items are the hard items in the EASY test but the easy items in the HARD test".

19. These are the general features associated with myRasch software. There is only one thing worth mentioning is that myRasch does not provide the function of test equating. The interested readers may consult the specific documentation or support resources for myRasch to get detailed and up-to-date information about its capabilities.
20. For more information on Lexile measures, visit www. Lexile.com.
21. Everett V. Smith, Jr. Professor of Educational Psychology at the University of Illinois at Chicago, Associate Editor of JAM, and member of the Editorial Committee for Educational and Psychological Measurement and Journal of Nursing Measurement. Specializing in psychological measurement, with a particular focus on Rasch measurement, his research interests and expertise include the design of tests and scoring scales, the analysis of potential construct measurements and the robustness of testing models, the application of Rasch measurement in social, behavioral, health, rehabilitation, and medical sciences, involving standard reference evaluation and norm reference evaluation. Authored works include Rasch Measurement: Theory, Models, and Applications (2004); Rasch Measurement: Advanced and Specialized Applications (2007), and Criterion-Reference Testing: Practice Analysis to Score Reporting Using Rasch Measurement Models (2009).
22. Richard M. Smith, Senior Director of Data Recognition Corporation and Founding Editor of Rasch Measurement's authoritative journal, *Journal of Applied Measurement* (JAM). Dr. Smith developed tests for the American College Test (ACT), guided the admission plan for dental and optometry tests of the American Dental Association, and conducted psychometric analysis for large-scale assessment programs such as the California High School Graduation Examination (CAHSEE), Golden State Examination, End of Course Examination Program (EOCEP) and PASS program in South Carolina, as well as the Alaska Standard Based Assessment. Apart from this, Dr. Smith has also conducted in-depth research on the Rasch measurement model and has published numerous related works. He has also served as a lecturer for multiple professional training sessions for professional organizations such as the American Educational Research Association (AERA) and the American Public Health Association (APHA). Many books on applied measurement have been published through JAM Press under his leadership.
23. This section is mainly based on the keynote speech "Rasch in Malaysia: A brief history, Challenges, and a Peek into Future delivered by Dr. Mohd Nor Mohd Zali who is the vice chairman of PROMS at PROMS 2017, Kota Kinabalu, Sabah, Malaysia.
24. The original book is in Malay, i.e., Asas Model Pengukuran Rasch
25. The original book is in Malay, i.e., Applikasi Pemodelan Rasch
26. For more details, interested readers may refer to RMT. Fall 2024, 37(1), ISSN 1051-0796.
27. The Benjamin Drake Wright Senior Scholar Award is a prestigious accolade given by the Rasch Special Interest Group (SIG) of the AERA. This award is presented biannually to recognize outstanding contributions to the field of Rasch measurement. Named in honor of Benjamin Drake Wright, a prominent figure in the development and application of Rasch measurement theory, the award acknowledges senior scholars who have made significant advancements in

research, theory, or practice related to Rasch models. Recipients of this award are typically individuals who have demonstrated a sustained impact on the field through their scholarly work, leadership, and mentorship.

28. At the conference considerable attention has been given to the roles of educational measurement, evaluation and assessment with a view to improving the education systems in general and to adequately prepare the young generation to meet the ever growing demands of the 21st century in particular. The conference is aimed at those working in all sectors of education and training who have a keen interest in advancing their knowledge and skills in educational measurement, evaluation, and assessment.
29. In Zhang & Hu (2000). PC-BILOG and Its Application in China: Equating and Item-banking. In Jose Lai & Pauline Po-yee Tam (Eds.), Crosslink in English Language Teaching. Vol. 1. English Language Teaching Unit. The Chinese University of Hong Kong. ISSN 1605–7511.
30. For details, see Muraki, E., & Bock, R.D. (1997). PARSCALE: IRT Item Analysis and Test Scoring for Rating-scale Data. p.146. SSI International. ISBN: 0-89498-042-4. http://www.ssicentral.com

Bibliography

Abd Razak, N., & Abdullah, W.M.R. (2003). Amalan Pengurusan Pengetua Sekolah dari Perspektif Guru (School Principals' Management Practices, from Teachers' Perspectives), Diges Pendidik Jilid 3, Bil. 1/2003, USM.

Abdullah, M.E. (2004). Mengukur Keberkesanan Model Konstruk Pengetahuan Berstruktur (KPB) dalam Pengajaran dan Pembelajaran Biologi dengan Menggunakan Model Pengukuran Rasch (Measuring Effectiveness of Structured Knowledge Construct Model in Teaching and Learning of Biology, using Rasch Measurement Model), Diges Pendidik Jilid 4, Bil. 2/2004, USM.

Abdullah, W.M.R. (1989). One Is Sufficient or Two Are Necessary: A Substitute for Clarifying Variables of Teaching through Factor Analysis of Rasch Standardized Residuals, Annual Meeting of the American Educational Research Association, San Francisco.

Abu Kasim, N.L. (2007a). Using the Rasch Measurement Model for Standard Setting of the English Language Placement Test at the IIUM, PhD Thesis. Malaysia: USM.

Abu Kasim, N.L. (2007b). Exploring Rater Judging Behaviour Using the Many-Facet Rasch Model, Faculty of Communication and Modern Languages. Malaysia: Universiti Utara.

Andrich, D., & Marais, I. (2019). A Course in Rasch Measurement Theory. Singapore: Springer.

Andrich, D., Sheridan, B., & Guangzhong, L. (2001). RUMM2030 Getting Started Manual.Copyright@1997-2012 Windows XP/Win7 Rumm Laboratory Pty. Ltd., Sydney, Australia.

Angoff, W.H. (1984). Scale, Norms, and Equivalent Scores. USA: Educational Testing Service.

Attali, Y., Saldivia, L., Jackson, C., Schuppan, F., & Wanamaker, W. (2014). Estimating item difficulty with comparative judgments. ETS Research Report Series, 1–8. https://doi.org/10.1002/ets2.12042

Aziz, A., Mohamed, A., Zaharim, A., Zakaria, S., Ghulman, H.M., & Masodi, M.S. (2008). Evaluation of Information Professionals Competency Face Validity Test Using Rasch Model. Proceedings of the 5th WSEAS/IASME International Conference on Engineering.

Bachman, L.F. (2010). Foreword. In L. Cheng & A. Curtis (Eds.), English Language Assessment and Chinese Learner (pp. x–xii). New York and London: Routledge.

Baird, J.-A. (2007). Alternative Conceptions of Comparability. In P. E. Newton, J. Baird, H. Goldstein, H. Patrick, & P. Tymms (Eds.), Techniques for Monitoring the Comparability of Examination Standards (pp. 124–156). London: Qualifications and Curriculum Authority.

Baker, F.B. (2001). The Basics of Item Response Theory. College Park, MD, USA: ERIC Clearinghouse on Assessment and Evaluation/University of Maryland.

Baker, F.B., & Kim, S.-H. (2004). Item Response Theory. Parameter Estimation Techniques. New York and London: CRC Press Taylor & Francis Group.

Benton, T. (2020). How Useful Is Comparative Judgment of Item Difficulty for Standard Maintaining? Res. Matters: Cambridge Assess. Publication 29, 27–35. Cambridge, UK: Cambridge Assessment.

Benton, T. (2021). Comparative judgment for linking two existing scales. Frontiers in Education, 6, 775203. https://doi.org/10.3389/feduc.2021.775203

Bond, T., & Fox, C.M. (2015). Applying the Rasch Model. In Fundamental Measurement in the Human Sciences. New York: Taylor & Francis Group.

Bond, T., & Yan, Z. (2018). Exporting to China: The Future of a Genuine Collaboration with the West. In Q. Zhang (Ed.), Proceedings of Pacific-Rim Objective Measurement Symposium (PROMS 2016). Springer Nature. https://doi.org/10.10079/978-981-10-8138_5

Bond, T., Yan, Z., & Heene, M. (2021). Applying the Rasch Model. Fundamental Measurement in the Human Sciences (4th Ed). New York: Routledge.

Bramley, T. (2005). A ordering method for equating tests by expert judgment. Journal of Applied Measurement, 6(2), 202–223. https://doi.org/10.1080/02671522.2010.498147

Din, R., Ahmad, M., & Sidek, N.M. (2009). Kesahan dan kebolehpercayaan soal selidik gaya e-pembelajaran (else) versi 8.1 menggunakan Model Pengukuran Rasch (Validation of questionnaire on E-Learning styles Version 8.1 using Rasch Measurement Model). Journal of Quality Measurement and Analysis, 5(2), 81–93.

Ehninger, J., Knigge, J., & Rolle, C. (2022). Why certain items more difficult than others in a competency test for music related argument? Frontiers in Education, 7, 1013841. https://www.frontiersin.org/articles/10.3389/feduc.2022.1013841/abstract

Fischer, G.H., & Molenaar, I.W. (1995). Rasch Models: Foundations, Recent Developments, and Applications. New York: Springer.

Hambleton, R., & Swaminathan, H. (1985). Item Response Theory: Principles and Applications. Dordrecht, The Netherlands: Kluwer Academic Publisher.

Hambleton, R.K. (1989). Principles and Selected Applications of Item Response Theory. In R.L. Linn (Eds.), Educational Assessment (3rd Ed, pp. 147–200). Washington DC, USA: Sponsored jointly by the National Council on Measurement in Education and the American Council on Education. ISBN 0-8268-9751-5

Holland, P.W., & Rubin, D.B. (1982). Test Equating. New York: Academic.

Kane, M.T. (2010). Validity and fairness. Language Testing, 27(2), 177–182. http://ltj.sagepub.com/content/27/2/177.abstract

Kolen, M.J., & Brenman, R.L. (1995). Test Equating: Methods and Practices. Berlin, London and New York: Springer-Verlag, Inc. Springer Series in Statistics. ISBN 0-387-94486-9.

Kolen, M.J., & Brenman, R.L. (2004). Test Equating, Scaling, and Linking Methods and Practices. Berlin, London and New York: Springer-Verlag, Inc. Springer Series in Statistics. ISBN 978-1-4757-4310-4.

Linacre, J.M. (2005). Rasch dichotomous model vs. one-parameter logistic model. Rasch Measurement Transactions, 19(3), 1032.

Linacre, J.M. (2016). A User's Guide to WINSTEPS Rasch-Model Computer Programs: Program Manual 3.92.0. ISBN 0-941938-03-4.

Livingston, S.A. (2004). Equating Test Scores (without IRT). USA: Educational Testing Service.

Lord, F.M. (1980). Applications of Item Response Theory to Practical Testing Problems. Mahwah, NJ, USA: Erlbaum.

Mahzir, A.M. (2008). The Application of Rasch Analysis. In The Development of The Malaysian H.E. Critical Thinking Instrument (MACTI): Conceptualisation and Psychometric Properties, Rasch Measurement Conference, Perth: University of Western Australia.

Mislevy, R.J. (1992). Linking Educational Assessments: Concepts, Issues, Methods, and Prospects. Princeton, NJ: ETS Policy Information Center.

Mislevy, R.J., & Bock, R.D. (1986). PC-BILOG: Item Analysis and Test Scoring with Binary Logistic Models. [Computer Software]. Scientific Software, Moorsville, IN, USA.

Mislevy, R.J., & Bock, R.D. (1990). PC-BILOG: Item Analysis and Test Scoring with Binary Logistic Models. [Computer Software] (2nd Ed). Scientific Software, Moorsville, IN, USA.

Mok, M.M.C., & Zhang, Q. (2014). Constructing Variables. Book of Abstracts Volume (I). Maple Grove, MN, USA: JAM Press. ISBN 978-1-934116-11-1

Mok, M.M.C., & Zhang, Q. (2015). Constructing Variables. Book of Abstracts Volume (II). Maple Grove, MN, USA: JAM Press. ISBN 978-1-934116-10-4

Mok, M.M.C., & Zhang, Q. (2018). Introduction to Rasch Measurement. ISBN 978-1-934116-13-5. JAM Press. Maple Grove, MN, USA.

Noguchi, H.(1986). An equating method for latent trait scales using common subjects' item response patterns. Japanese Journal of Educational Psychology, 34(4), 43–50.

Panayides, P., Robinson, C., & Tymms, P. (2010). The assessment revolution that has passed England by: Rasch measurement. British Educational Research Journal, 36, 611–626. https://doi.org/10.1080/01411920903018182

Samsudin, M.A., Osman, K., & Halim L. (2006). Application of Rasch Model in Classroom: Tracking the Growth of Student Understanding towards Physics Concepts in Different Problem-based Learning Environments. Petaling Jaya, Malaysia: Kuala Lumpur International Conference on Assessment (KLICA).

Sivakumaran, P.K. (2007). Further Methods of Correlations. In S.C. Gupta (Ed.), Fundamentals of Applied Statistics. New Delhi, India: Sultan Chand and Sons.
Steiner, S. (2021). "Better Measurement and Fewer Parameters! The True Value of Rasch over IRT", The keynote speech at PROMS2021, Nanjing, China.
Stenner, J. (2013). The Lexile Framework for Reading. Pre-Conference Workshop, PROMS2013, National Sun Yat-sen University, Kaohsiung, Taiwan, China.
Tian, Q.-Y. (2006). Analysis of Test Equating Improvement for HSK. Beijing: HSK Centre, Beijing Language and Culture University
Uysal, I., Kılıç, A. F., & Keleciolu, H. (2017). Use of Common Subjects in Test Equating: A Simulation Study. The 6th International Conference on Education (IC-ED-2017).
Wei, Y., & Morgan, R. (2016). An Evaluation of the Single-Group Growth Model as an Alternative to Common-Item Equating. Princeton: Educational Testing Service. NJ ETS Research Report Series ISSN 2330-8516. https://doi.org/10.1002/ets2.12087
Wilson, M., & Sloane, K. (2000). From principles to practice: An embedded assessment system. Applied Measurement in Education, 13, 181–208.
Wright, B. (1989). MESA Psychometric Laboratory. Retrieved from https://www.rasch.org/rmt/rmt34g.htm.
Wright, B.D. (1992). IRT in the 1990s: Which models work best? Rasch Measurement Transactions, 6(1), 196–200.
Wright, B.D., Mead, R.J., & Bell, S.R. (1980). BICAL: Calibrating Items with the Rasch Model. Research Memorandum 23 C. Chicago: University, Department of Education.
Wright, B.D., & Stone, M.H. (1979). Best Test Design Rasch Measurement. MESA.
Wu, R.Y. (2020). The General English Proficiency Test in Taiwan: Past, Present, and Future. In Lily I-Wen Su, Cyril J. Weir, & Jessica R.W. Wu. (Eds.), English Language Proficiency Testing in Asia. A New Paradigm Bridging Global and Local Contexts. New York and London: Routledge and Taylor & Francis Group. ISBN 978.
Wu, J., & Zhang, Q. (2018). Test Equating with WINSTEPS and GITEST: Different but Alike. In Q. Zhang (Ed.), Pacific-Rim Objective Measurement Symposium (PROMS) 2016 Conference Proceeding. Springer Nature. ISBN 978-981-10-8137-8. https://doi.org/10.1007/978-981-10-8138-5_6
Zhang, Q. (2011). Towards Better Interaction between Testing and Teaching. Keynote Speech Given at the 5th National TEFL/1st Mongolia TESOL Conference, Ulaanbaatar, Mongolia, Oct 7–9, 2011. Mongolia.
Zhang, Q. (2012). Towards International Practice of Language Testing in China. Keynote Speech Given at the PROMS2012, Jiaxing, China, August 6–9, 2012.
Zhang, Q. (2019). Rasch Model: Retrospect and Prospect in China. In Myint Swe Khine (Ed.), International Trends in Educational Assessment. Emerging Issues and Practices. Leiden and Boston: BRILL SENSE. Retrieved from http://catalog.loc.gov.
Zhang, Q. (2021a). Rasch-GZ, The First Chinese Version of Rasch-Based Item Analysis and Test Equating. In Department of Psychology, Nanjing Normal University (Chairs), Online Symposium. Online Symposium. PROMS2021, Nanjing, China. Retrieved from http://www.proms.promsociety.org
Zhang, Q. (2021b). Rasch Model and Test Equating in China: Rasch Model in Chinese Version Has Come. The 6th International Conference on Language Testing and Assessment. Shanghai, China.
Zhang, Q. (2022a). From GITEST to RASCH-GZ: Inheritance and Development of Rasch-based Research in China. http://dx.doi.org/10.20431/2349-0381.09S1001

Zhang, Q. (2022b). Rasch Model and Test Equating in China: The Most Updated in Mariam Haji Monek (Chairs), International Education e-Colloquium, the 22nd-24th Feb. 2022. KL, Malaysia.

Zhang, Q. (2023). RASCH-GZ (Version 2.0) [Computer software]. http://www.rasch-gz.com

Zhang, Q. (2024a). Rasch-GZ: The Most Updated Research Development in China. In Q. Zhang (Ed.). Proceedings of Pacific-Rim Objective Measurement Symposium (PROMS2023) AP/Springer Nature. https://doi.org/10.2991/978-94-646-494-5_22

Zhang, Q. (Ed.). (2024b). Proceedings of Pacific-Rim Objective Measurement Symposium (PROMS 2023) AP/Springer Nature. https://atlantis-press.com/proceedins/proms-23

Zhang, Q. (2004c). BILOG and PARSCALE: Different yet Alike. In Jayanti Banerjee & Dianne Wall (Eds.), *Language Testing Update (LTU)* Issue 35-Spring 2004. Centre for Research in Language Education, Lancaster University, England: International Language Testing Association Official Newsletter.

Zhang, Q., & Hu, X.X. (2000). PC-BILOG and Its Application in China: Equating and Item-banking. In Jose Lai & Pauline Po-yee Tam (Eds.), Crosslink in English Language Teaching. Vol. 1. HK, China: English Language Teaching Unit.

黄晓婷等编译. (2019). 《基于建构理论的量表设计》. 湖南教育出版社. 国家社会科学基金重大项目 (17ZDA323) 核心成果

张知干(主编). (2021). p. 270. 《广东省优秀社会科学家传略》中山大学出版社.

刘育明. (2020). [测验等值量表制订和联结的方法与实践]. (第三版) ISBN 978-7-5446-5996-5 上海外语教育出版社.原作: Michael J. Kolen & Robert L. Brennan. (2004). Test Equating, Scaling, and Linking Methods and Practices. Springer-Verlag New York, Inc. Springer Series in Statistics. ISBN 978-1-4757-4310-4.

金艳、杨惠中. (2018). 走中国特色的语言测试道路: 大学英语四、六级考试三十年的启示. 外语界, (2) 29–39

桂诗春、李崴、张权. (1993). 项目反应理论在高考成绩等值上的应用. 载国家教委考试中心主编第四届全国教育考试科研讨论会论文选编 (pp. 391–393)。中国和平出版社, 北京. ISBN 7-80101-089-2/C.64.

桂诗春. (1991). 项目反应理论在高考试题等值上的应用 《中国考试》 北京. ISSN 1005-8427.

桂诗春. (2007). 项目反应理论在高考英语科等值中的应用. 载桂诗春 (主编). 《桂诗春英语教育自选集》 (pp. 433–440). 外语教学与研究出版社.

桂诗春. (1990) 题库建设讲话(二). [J]. 现代外语. (4), 66–72. 1990 (1): 66–72,59.

李崴. (2000). 《高考改革探索与实践》 高等教育出版社.

李筱菊. (1997). 《语言测试的艺术与技术》 湖南教育出版社.

谢小庆. (2022). Equating Handouts. wenku.baidu.com

谢小庆. (1998). Experimental Study on HSK Equating. Teaching Chinese in the World. Issue 3. 1998.

朱正才、杨惠中、杨浩然. (2003). Rasch 模型在 CET 考试分数等值中的应用, 现代外语, (1): 69–75.

张权. (2008). 《结构方程建模在语言测试中的研究与应用》 高等教育出版社, 北京 ISBN 978-7-04-024640-7

张权. (2004). 《项目分析与等值技术在语言测试中的研究与应用》 高等教育出版社, 北京

国家教委考试中心统计分析与技术处. (1993). 试论 CTT 分数等值与样本的关系. 载国家教委考试中心主编. 《第4届全国教育科研讨论会》 论文选编.

国家教委考试中心统计分析与技术处. (1993). IRT 题目参数的估算与等值. 载国家教委考试中心主编. 《第4届全国教育科研讨论会》 论文选编.

4

RASCH-GZ: Operating Steps for CTT-based Item Analysis

This chapter provides a user guide to RASCH-GZ, the most updated Rasch-based software and platform for both CCT-based Item Analysis and Rasch-based Test Equating successfully developed over the past three years (2019–2021). The present chapter can also be used as handout for one-day workshop run by the program developers, focusing on CTT-based item analyses and the Rasch-based test equating (the latter will be elaborated in Chapter 5). All the content in the present chapter is based on pre-conference workshops conducted both online and on site of Pacific-Rim Objective Measurement Symposium (PROMS) conferences in Malaysia, Thailand, China Macau and universities and colleges across China Mainland since 2019.

4.1 Features of the Rasch-GZ

Technically, the Rasch-GZ used Delphi, JAVA, and Python. Delphi has played its advantages in developing Windows desktop application system, which has realized the interface for visual data editing fully compatible with Excel, thus making easier data editing in Excel and data importing into the system for analysis, i.e., users can process big data matrix, i.e., unlimited items by unlimited subjects. At the same time, desktop applications have excellent support of Chinese characters. The purpose is to provide visualization, networking, both English and Chinese language support, and the Window system suitable for users. Python has its unique advantages in data processing, and JAVA in networking, which is stable and reliable.[1]

The Rasch-GZ has two functions: Item analysis and test equating. The item analysis module is based on the classic test theory (CTT). This includes multidimensional analysis (see Section 2.1). Test equating module is based on Rasch model (see Section 3.7) (Zhang, 2023, 2022, 2021, 2012b). The Rasch-GZ adopts menu operating, i.e., the steps for the Rasch-GZ to getting started are straightforward and no command file or control file to be created. Furthermore, data editing does not need to be as cumbersome as the initial software development. Output files can be in WORD or PDF form and can be in English or Chinese. Plotting is applicable and online technical support is available.

DOI: 10.1201/9781003665595-4

Apart from this, incorporated with AI, Rasch-GZ is capable of popping out hints either in spoken mode or in written mode on the computer screen while data is being edited. Additionally, the users may also download the short video guide, the most intuitive manual, prior to getting started so as to make sure you do know how to implement them correctly. So watch it and get to work!

4.2 Installation and Startup

Installation instructions are included at the Rasch-GZ website,[2] so download and install the Rasch-GZ now on your laptop or desktop PC. The free online student version is preloaded with the data files (30 x 40 sample size) to confirm that you know how to implement them correctly. Now let's usher you, step by step, from getting started to the interpreting of each analysis report Figure 4.1.

FIGURE 4.1
The logo of Rasch-GZ

4.3 System Login to the RASCH-GZ

To get started, let's click on the program icon (see the right) to obtain the login homepage of Rasch GZ interface on the computer screen as shown in Figure 4.2.

FIGURE 4.2
The starting interface page of Rasch-GZ

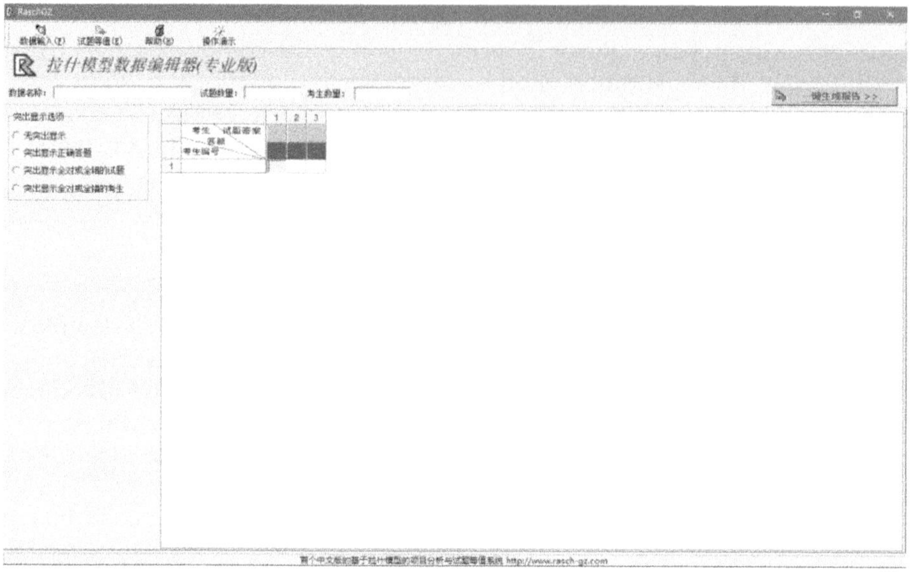

FIGURE 4.3
The data import interface page of Rasch-GZ

Like any other software currently used, we follow the four steps specified below to log in:

Step 1: Enter username/password (password set by the user themselves).

Step 2: If you need help, please click the "Help" button.

Step 3: After logging in, if you want to exit, please click the "Exit" button.

Step 4: Click the "Login" to enter the spreadsheet interface, as shown in Figure 4.3.

4.4 Data Input/Editing

RASCH-GZ data can be input either by hand entry or data file import. For data input by hand entry, the first step is constructing the data file on the spreadsheet. All you need to do is entering the data file directly according to the data format. If you encounter any problems, the system will automatically prompt a window, providing guidance information in a timely manner either in written or spoken mode to guide users, step by step to complete the data file for analysis.

4.4.1 Data Type and Format

The Rasch-GZ accepts text file, EXCEL file, CSV, and so on. The data type can be of chars like ABCDACDABACD. In the case of the raw data in chars, the corresponding key to each test item will be input in the 3rd line on the spreadsheet (as shown in Figure 4.4). In the case of integers such as 101010101111100 wherein "1" represents a correct answer, while "0", an incorrect answer. No key answers are needed. The data matrix of 30-item by 40-subject on the spreadsheet is illustrated in Figure 4.4 where the raw data is in chars, and the indication for each part in the data file is given. Needless to say, in today's data analysis, the common practice remains that a data matrix is of a rectangular ASCI file wherein cases are entered horizontally and items, vertically, with each cell, e.g., in Excel format, containing just one ASCI character, either a char or an integer. The Rasch-GZ follows these same requirements.

4.4.2 Keyboard Input

To create data files manually, we just follow the example of data matrix illustrated in Figure 4.4. Once done, the current data file is saved. The Rasch-GZ will pop up a window, indicating that the user has successfully saved the

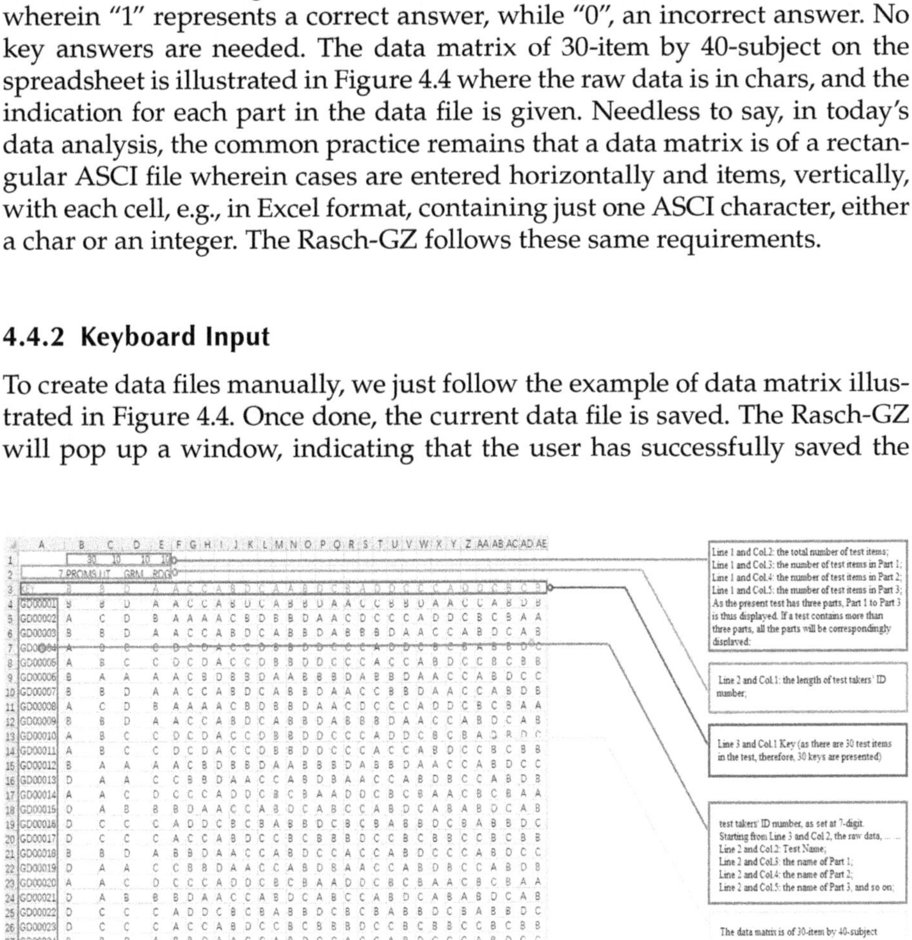

FIGURE 4.4
The data import format required by Rasch-GZ

data with the file name of "GD08" defined by the user. And the system automatically adds "dat", the extension to "GD08.dat", indicating this is a data file. This is good for portability in case you need data transfer or data reproducibility for paper writing and research reports.

In doing so, users needn't worry about any mistakes keyed in accidently because in the case of the Rasch-GZ, should anything (or any char, integer or any cell) go wrong, the system will pop out a hint suggesting how to correct meanwhile the place where correction is to be made is in highlight on the spreadsheet.

where

Starting from

Line 1 and **Col 2**: 30, the total number of test items

Line 1 and **Col 3**: 10, the number of test items in Part 1.

Line 1 and **Col 4**: 10, the number of test items in Part 2.

Line 1 and **Col 5**: 10, the number of test items in Part 3.

As the present test has three parts, Parts 1–3 are thus displayed. In a test that contains more than three parts, all the parts will be correspondingly displayed.

Line 2 and **Col 1**: The length of test takers' ID number.

Here, 7 indicates the ID number is 7 digits.

Line 2 and **Col 2**: Test Name.

Line 2 and **Col 3**: The name of Part 1.

Line 2 and **Col 4**: The name of Part 2.

Line 2 and **Col 5**: The name of Part 3.

Line 3 and **Col 1**: Key. As there are 30 test items in the test, therefore, 30 keys are input herein.

Line 4 and **Col 2**: The raw data … …

In the case of integers such as 101010101111100, wherein "1" represents a correct answer, while "0", an incorrect answer. No key answers are needed. The above data matrix of 30 items by 40 subjects would be input like this. Figure 4.5 shows the segment of the data file where rows are cases; columns, items; and no empty cells.

In Figure 4.5, only specified are the first two lines, respectively, indicating the total number of the test, the number of test items in each part, and the test takers' ID number, the test name, and the name of each section. "X" is used for any missing data of whatever in single cell.

```
                30    10    10    10
          7   PROMS   LIT   GRM   RDG
GD00001 11111111110110101101010111110
GD00002 1111111110111111111010 0111111
GD00003 111111111111111111111110011011X
GD00004 011111110111111011111110111111
GD00005 011111110111111011111110111111
GD00006 111111111111111101011110111111
GD00007 111110111110111110111110111110
GD00008 111111111011011111101101111111
GD00009 111010011111111111011110101011
GD00010 01111111111X111111111110111110
        ... ... ... ...
GD00038 000000000000000000000000000000
GD00039 111111111111111111110111111011
GD00040 111111111110111111101110110111
```

FIGURE 4.5
The sample data matrix of Rasch-GZ

4.4.3 Special Function in Data Checking

Apart from this, the Rasch-GZ has special functions to automatically check and delete invalid cases and items in data set. This includes to check and delete:

(1) A case that has got correct answers to all the items;
(2) A case that has got incorrect answers to all the items;
(3) An item that all the test takers answered correctly;
(4) An item that all the test takers answered incorrectly;
(5) Two or more cases turn out to be exactly the same, i.e., possible cheating.

As we know, the first four cases (1)–(4) would make the program abort in the case of Maximum Likelihood (ML) is used. In such cases, the users using other software programs have to quit the program, enter the data file to check each case. This is hassle and time-consuming for using both Rasch-based and IRT-based software packages. The Rasch-GZ has put this into consideration, trying to be all things to all Rasch analysts. Figure 4.6 shows the

```
                 30     10     10     10
       7     PROMS    LIT    GRM    RDG
GD00001 111111111101101011010101111110
GD00002 1111111110111111111010 0111111
GD00003 11111111111111111111110011011X
GD00004 011111101111110111111101111111
GD00005 011111101111110111111101111111
GD00006 111111111111111101011110111111
GD00007 111110111101111101111101111110
GD00008 111111111011011111101101111111
GD00009 111010011111111111011110101011
GD00010 0111111111111X11171111110111110
             ... ... ... ...
~~GD00038 000000000000000000000000000000~~
~~GD00039 111111111111111111111111111111~~
GD00040 111111111111011111110111011011
```

FIGURE 4.6
The data sample showing all the correct or/and incorrect answers will be automatically deleted prior to calibration by Rasch-GZ

```
                 30     10     10     10
       7     PROMS    LIT    GRM    RDG
GD00001 111111111101101011010101111110
GD00002 1111111110111111111010 0111111
GD00003 11111111111111111111110011011X
~~GD00004 011111101111110111111101111111~~
~~GD00005 011111101111110111111101111111~~
GD00006 111111111111111101011110111111
GD00007 111110111101111101111101111110
GD00008 111111111011011111101101111111
GD00009 111010011111111111011110101011
GD00010 0111111111111X11171111110111110
             ... ... ... ...
GD00040 111111111111011111110111011011
```

FIGURE 4.7
The data sample showing all the same response patterns will be automatically deleted by Rasch-GZ to prevent possible cheating

segment of the data file where (1)–(2) and (3)–(4) were checked and deleted. Figure 4.7 shows the segment of the data file where possible cheating cases were checked and deleted.

4.4.4 External Files Import

To import external data, the user needs to have a data file edited in Rasch-GZ format (as shown in Figure 4.4).

To import text data, click "Import GITEST Data" from the "Data Input" menu in the upper left corner of the menu bar, and the following dialog box will pop up as seen in Figure 4.8. By now, the user simply selects the corresponding GITEST format data file and opens the current data window to import data.

To import EXCEL data, click "Import EXCEL Data" from the "Data Input" menu in the menu bar. The dialog box will appear (as seen in Figure 4.9). We select the corresponding EXCEL format data file and open the current data editing window to import data. The Rasch-GZ accepts older versions as well as versions after 2007.

To use the Rasch-GZ to do any analysis, up to now your job is basically done because the essential part using the Rasch-GZ to do any analysis is simply constructing data file. **No command file or control file is needed.**

What to do next is to find, in the menu on the upper left corner of the menu bar, which analysis users plan to conduct. With just one click, the user could obtain the output file of analysis they need.

FIGURE 4.8
GITEST external data file import dialog

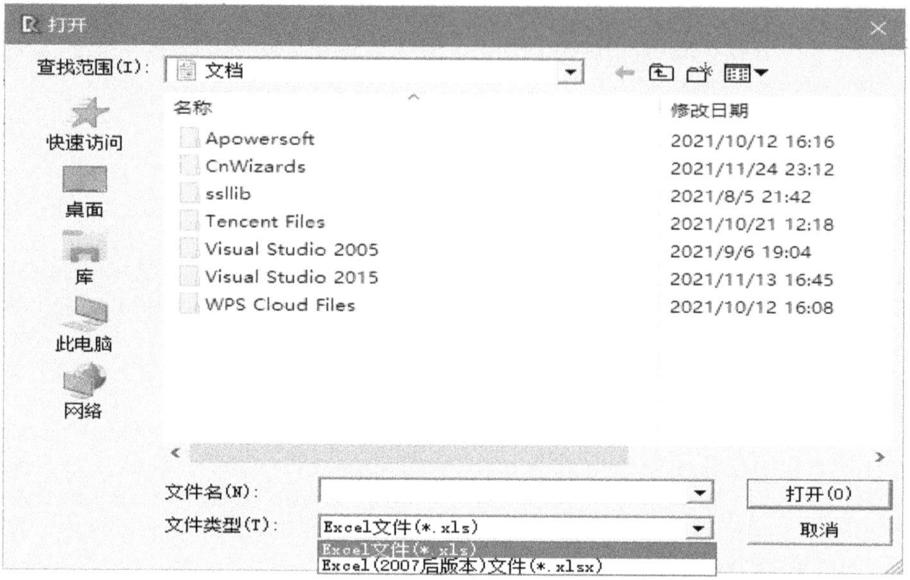

FIGURE 4.9
EXCEL external data file import dialog

4.5 The Rasch-GZ: First Try

Now, we are going to take you through the first analysis and interpretation of CTT-based item analysis. Once you've gone through successfully, we are sure that you could definitely repeat the same steps with your own file for any dichotomous data analysis.

4.6 Operating Steps for CTT-based Item Analysis

This is the simple way to begin. Figure 4.10 is the interface of the CTT-based Item Analysis of the Rasch GZ system.

Once the data file is imported successfully, or manually input is completed, we could obtain the required item analysis output reports simply by clicking of mouse. To repeat, Rasch GZ users do not need to create any command files or control files as required by most other software packages (see Section 3.8). Instead, users just get familiar with the simple menu operations like using a point-and-shoot camera. In case the data matrix is much larger, a progressing bar will prompt on the screen while the data is being processed. After

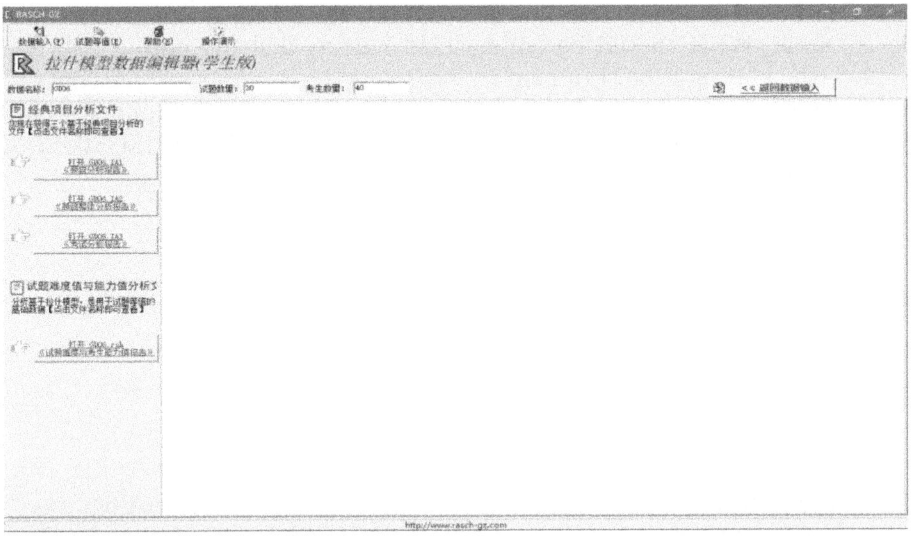

FIGURE 4.10
The RASCH-GZ interface for item analysis

it completes, the interface for item analysis output report will appear. For example, if the user specified and saved the data file name as "**RGZ001.dat**", there would be four output file names generated with different extensions in the left column as shown in Figure 4.11.

Sometimes, the user only needs one specific output data file according to his own research purpose. In such a case, just click the required data analysis report. These four important output data files are generated from

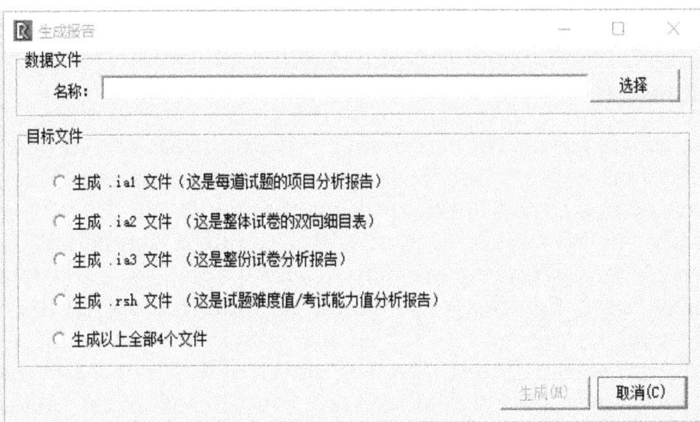

FIGURE 4.11
The interface for four files to be generated

the CTT-based module. They are RGZ001.ia1, RGZ001.ia2, RGZ001.ia3, and RGZ001.rsh. What should be emphasized here is that the output data file with "-rsh" ending is NOT of item analysis. It is the Rasch-based file containing item difficulty parameters and ability estimates to be used for test equating. In what follows, all the four output data files: RGZ 001.**ia1,** RGZ001.**ia2,** RGZ001.**ia3,** and RGZ001.**rsh** will be interpreted one by one.

4.7 Output Data Files

As indicated in Figure 4.11,

(1) RGZ 001.**ia1** is "Item analysis report for each test item";
(2) RGZ 001.**ia2 is** "two-way checklist table (TWCT) about difficulty-discrimination of a test paper with detailed information about each section of a test paper";
(3) RGZ 001.**ia3** is "An analysis report about a whole test paper". These three files are based on the CTT, with the extension -.ia1, -.ia2, and -.ia3, respectively;
(4) RGZ 001.**rsh** is the data file containing all test item difficulty parameters and test takers' ability estimates based on **one test paper.** In the Rasch-GZ, data files with –rsh ending are indispensable files for conducting test equating. Therefore, to conduct test equating, at least **two** data file with "-rsh" ending are needed. To be more exact, we need two data file with "-rsh" ending, one is based on the basal test and the other is based on the test to be equated. With these two "-rsh" ending files, the test equating is feasible.

4.7.1 RGZ001.ia1

Figure 4.12 shows part of the screenshot of RGZ001.**ia1.** The complete output file of RGZ001.ia1 is presented in Appendix III with detailed analysis report for each test item at the end of the present chapter.

Technically, the Rsach-GZ system will generate a detailed item analysis report for each corresponding test item in a test paper, e.g., if the data file contains 100 test items, the Rasch-GZ would generate 100 output detailed item analysis for each test item. In CTT, "-ia1 file" is very important for professionals to check how well each test item was designed and to use as a reference to moderate, or to reject poorly designed test items based on raw data collected from a pre-test administered to the target subjects conducted.

In what follows, we pick up two representative cases from Appendix III at the end of the chapter to illustrate the ideas addressed about the item analysis

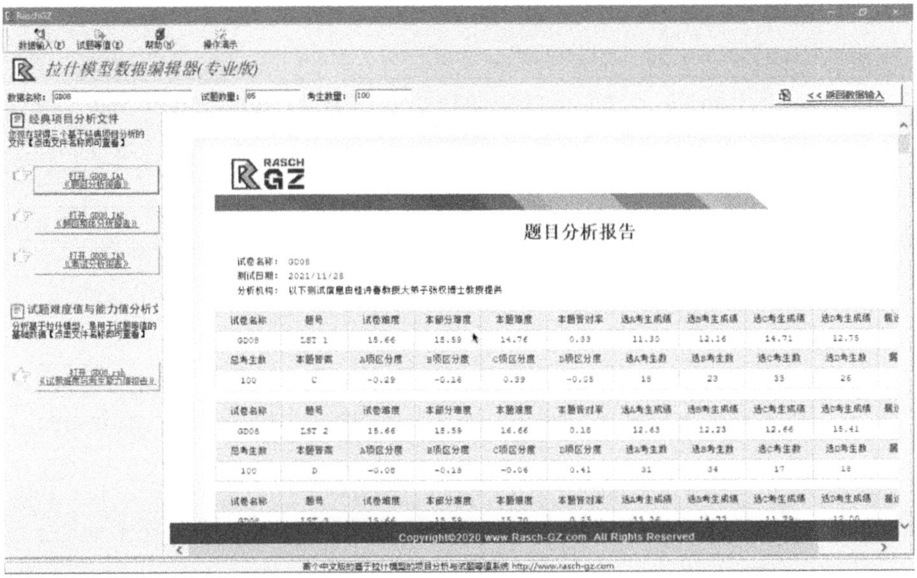

FIGURE 4.12
Part of sample file ending with ia1

in (Sections 2.1–2.1.13, Chapter 2). Let's look at the first two items presented in Tables 4.1 and 4.2.[3]

(1) As can be seen in Table 4.1,
The first line indicates in order:

Test code, i.e., the test name is GD08;

Item No. LST 1 here indicates this is the first test item of listening comprehension part;

Pt refers to the average difficulty level of the whole test paper. Here, 11.06 is the Δ-value. As we know, the $\Delta = 13 + 4Z$, ranging from 1 to 25. The higher the value, the greater the difficulty, with the constant value being 13 indicating 50% of the total test takers got the correct answers (see Section 2.1.11); Now, the Δ-value being 11.06, implying this is an EASY test.

TABLE 4.1
The item analysis for the first test item of "RGZ001.ia1"

Test Code	ITEM No.	Pt	Pi	Pd	P	MA	MB	MC	MD	MO
GD08	LST 1	11.06	11.41	8.23	0.88	12.94	−14.72	14.10	12.41	−14.72
No. of CAN	**Key**	**Ar**	**Br**	**Cr**	**Dr**	**A**	**B**	**C**	**D**	**O**
43	A	−0.06	0.00	0.15	−0.07	38	0	3	2	0

TABLE 4.2
The item analysis for the second test item of "RGZ001.ia1"

Test Code	Item No.	Pt	Pi	Pd	P	MA	MB	MC	MD	MO
GD08	LST 2	11.06	11.41	11.69	0.63	14.54	13.66	12.42	10.71	−14.72
No. of CAN	**Key**	**Ar**	**Br**	**Cr**	**Dr**	**A**	**B**	**C**	**D**	**O**
43	B	0.21	0.28	−0.07	−0.44	4	27	2	10	0

Similarly,

Pi is Δ = 11.41, i.e., the average difficulty of the listening comprehension part a little bit lower than 13; in other words, this part is also EASY.

Pd is Δ = 8.23 indicating this is an EASY item.

P refers to the P-value, here 0.88 indicates the overall correct rate is 88%.

MA refers to the total average score (12.94 here) obtained by the test takers who chose Option A, i.e., the correct answer.

MB refers to the total average score (−14.72 here) obtained by the test takers who chose Option B.

MC refers to the total average score (14.10 here) obtained by the test takers who chose Option C.

MD refers to the total average score (12.41 here) obtained by the test takers who chose Option D.

MO refers to the test takers who did not choose any option.

(2) The second line indicates in order:

No. of CAN refers to the total number of the test takers, here is 43.

Key indicates A is the correct answer for this test item.

Ar refers to the discrimination index (DI; bi-serial) of Option A. Here is -0.06. Please refer to (Section 2.1.13), the discrimination for correct answers equal to or greater than 0.3 and less than 0.7 is fully accepted. Here, -0.06 indicates a poor item, implying that the discriminating power is very weak and that some test takers with lower scores happened to get this correct answer.

Br refers to the DI (bi-serial) of Option B. Here is -0.00, showing no discriminating power at all. Please refer to (2.1.13), any distractor must produce some distracting effect. DI of a distractor = 0, indicating a poor option. No distracting effect. Moderate or discard it because zeroed distractor is the root cause to decrease the difficulty of MC questions.

Cr refers to the DI (bi-serial) of Option C is 0.15. DI of a distractor greater than -0.3 and less than 0.2 is acceptable.

Dr refers to the DI (bi-serial) of Option D is -0.07. A weaker distractor but can be accepted.

There are two more points that need to be clarified here. At the first place, negative DI indicates that the test takers scoring high on the test may have answered the item incorrectly, while the test takers with low scoring got the correct answers. It is imperative to check up the original test item. This is often due to the poor wording in the test item stem or the other options, etc. Next, DI = 1, which is the ideal discrimination! This indicates that all the test takers scoring high in the test overall answered the item correctly. In contrast, all the low-scoring ones answer the item incorrectly. This can only be the explanation as the concept of DI in classroom teaching. In practical tests on large scale yet with high stakes, it is difficult to obtain such a test item with DI being 1 for the correct answer.

A refers to the number of test takers who chose Option A is 38. This shows us why this test item is easy with poor discriminating power because 38 out 43 test takers got the correct answer of this test item.

B refers to the number of test takers who chose Option B is 0. This shows us why the DI of this distractor is zeroed because no one took this option.

C refers to the number of test takers who chose Option C is 3.

D refers to the number of test takers who chose Option D is 2.

O refers to the number of test takers who did not choose any option is 0.

Conclusion: This test item needs major moderating, or discard it.

(3) Now let's check Table 4.2.

The first line indicates in order: Test code, Item No. Pt, and Pi, which are the same as the first item.

Pd is $\Delta = 11.69$ indicating this is an **Intermediate** item; the P-value is 0.63 indicating the overall correct rate is 63% (see Section 2.1.2);

MA shows the total average score (14.54 here) obtained by the test takers who chose Option A;

MB shows the total average score (13.66 here) obtained by the test takers who chose Option B, i.e., the key answer to the item;

MC shows the total average score (14.42 here) obtained by the test takers who chose Option C;

MD shows the total average score (10.71 here) obtained by the test takers who chose Option D;

MO shows that all the test takers made their own choices, and no one who gave up.

Starting from the second line, we could see in order:

No. of CAN shows totally 43 test takers sat for the test. **Key** answer is A for this test item. DIs of the four options are in order:

Ar is 0.21, which is acceptable for a distractor;

Br is 0.28, sort of weak for a key answer but can be accepted;

Cr is 0.07, which is acceptable for a distractor;

Dr is −0.44, a poorly designed distractor, and needs moderating;

A indicates there were 4 test takers who chose Option A, showing this distractor was working;

B indicates majority of the test takers, i.e., 27 out of 43 identified this option is the key answer. This shows us that this is an Intermediate item with good facility but sort of weak discrimination;

C indicates there were 2 test takers who chose this option, showing this option as a distractor was working;

D indicates there were 10 test takers who chose this option, showing this option as distractor is sort of strong and needs to be moderated appropriately;

O shows us no test takers gave up;

Conclusion: This test item can be accepted, given Option D as a distractor needs moderating.

4.7.2 RGZ001.ia2

In the Rasch-GZ, the file with "-ia2" ending is the second CTT-based item analysis output file. Figure 4.13 shows part of the screenshot of RGZ001.**ia2,** which is a two-way checklist table (TWCT) about difficulty-discrimination of all the test items within a test paper.

In this TWCT, the horizontal line indicates the P-value (0-1); vertical line shows the discrimination index (between 0.1 and 1). At the bottom of the table, clear at a glance are the expected quantity, the actual quantity, and the quantity that meets the requirements of P-value of the test items within a test paper, respectively, in a distribution. This can be summarized in the following three points:

(1) In some cases, though failing to meet the requirements, these calibrated test items are close to the required parameters, and the difficulty is appropriate. Putting the error factors into consideration, the test developers are supposed to retain these calibrated items for tests on small scale;

(2) It is convenient for testing professionals to adjust the item difficulty in a test paper. For example, if easier or more difficult items are observed, which makes the difficulty level of the test paper unbalanced, the solution is to add more difficult items and withdraw easier ones or vice versa. In this way, we could ensure that the overall difficulty of a test paper to meet the requirements;

(3) The replaced test items (up to requirements) can be properly saved for future use because these items have been well calibrated with good discriminative power.

On the whole, the TWCT generated by the Rasch-GZ gives a clear understanding of whether the test design conforms to the requirements that the test developers desired. In popular words, the adjustment of overall difficulty level of a test paper is simply selecting calibrated test items with more appropriate parameters to replace those that fail to meet the requirements. However, for most examination developers, for schools or social training centers, it is infeasible to set up an item bank. This is the ongoing situation that testing professionals are facing everywhere. Now with the Rasch-GZ, things are easy to handle in such a quantitative way. Appendix IV at the end of the chapter is a complete TWCT about difficulty-discrimination of a test paper with detailed information about each section of a test paper.

FIGURE 4.13
Part of sample file ending with ia2

4.7.3 RGZ001.ia3

In the Rasch-GZ, the file with "-ia3" ending is the third CTT-based item analysis output file. RGZ001.ia3 as shown in Table 4.3 is an output report for overall analysis about a whole test paper.

TABLE 4.3

A report of overall analysis of a test paper

Test Name: basal

Test date: 2025/06/03

The following report is generated via the Rasch-GZ

The test has a mean score of 27.01 having 27% of the answers correct, indicating it is a test of difficult level. The standard deviation is 5.29 against the expected standard deviation 16.67, having a range of 27, indicating the scores are centralized.

On the whole, the distribution of the scores is positively skewed. The discrimination index of the whole test paper turned out to be 0.16. The standard error of measurement is ±4.30. Factor analysis of the test shows that it accounts for 34.69% of the knowledge and skills of the examinees. It also shows:

LST, RDG belongs to a factor. Other factor(s) (<.4).

Item analysis shows that there are 1% of very easy items, 1% easy items, 22% items of medium level, 76% difficult items, 0% very difficult items. There are 66 items that do not meet the requirement.

Analysis of each component shows that LST has a mean score of 11.79, (against 40) having 29% of answers correct and 0.22 of rbis, indicating a section of difficult level. On the whole, in this LST component, there are 25 item(s) that do not meet the requirement.

GRM has a mean score of 7.65, (against 30) having 26% of answers correct and 0.23 of rbis, indicating a section of difficult level. On the whole, in this GRM component, there are 23 item(s) that do not meet the requirement.

RDG has a mean score of 7.57, (against 30) having 25% of answers correct and 0.25 of rbis, indicating a section of difficult level. On the whole, in this RDG component, there are 18 item(s) that do not meet the requirement.

For any question or doubt, please feel free to contact the following:

1. info@rasch-gz.com
2. qzhang141@rasch-gz.com

4.7.4 RGZ001.rsh

RGZ 001.rsh is the output file containing all the test item difficulty parameters and test takers' ability estimates based on one test paper. In the Rasch-GZ, output files with -rsh ending are indispensable files for conducting test equating. To conduct test equating, at least two output files with "-rsh" ending are needed. To be more exact, users need two data files with "-rsh" ending, one is used as the **basal test,** and the other is the test **to be equated**. With these two "-rsh" ending files, users can perform test equating. Appendix III at the end of the chapter presents all the test item difficulty parameters and test takers' ability estimates based on one test paper. We'll take it as the basal test, and the comparison of two sets of data will be discussed in Chapter 5.

4.8 Other Operations

This section will briefly introduce some other simple file operating of Rasch-GZ because in terms of basic menu and keyboard operations, Rasch GZ is basically the same as other software.

4.8.1 How to Open a Saved Data File

Click "Open" from the "Data Input" menu, and the "Open Data File" window will appear as shown in Figure 4.14.

4.8.2 How to Open Recently Used Data Files

Click "Open the recently used data" from the "Data Input" menu, and the file names of the 5 recently used data files would be displayed in the menu. Click the file name you need to get the data.

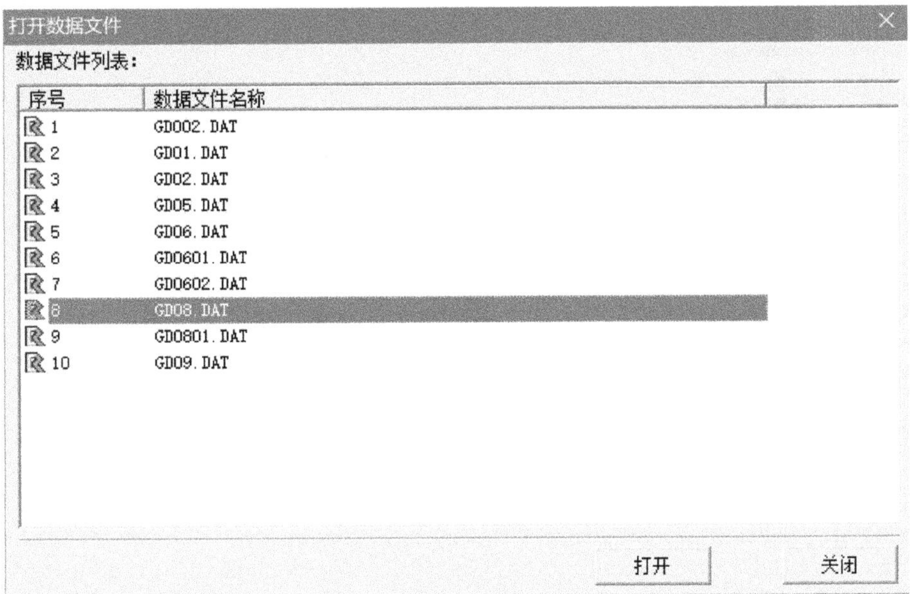

FIGURE 4.14
The data file folder of Rasch-GZ

4.8.3 How to Moderate the Property Value of the Data Currently Being Edited

Click "Properties" from the "Data Input" menu and the "Data Properties" window will appear. In Rasch-GZ, modifiable values refer to (1) data name, (2) number of test items, (3) number of test takers, (4) ID length of test takers, and (5) name and number of each section of the test paper.

By entering the new corresponding attribute value in the corresponding input box, the user can add or delete operations in the test section. Click the "OK" button to apply the modified properties, click the "Cancel" button to discard the modification.

4.8.4 Close the Current Data

Click "Close" from the "Data Input" menu to close the data currently being edited.

4.8.5 Save the Current Data as

Click "Save As" from the "Data Input" menu, and the "Data Properties" setting window will pop up to modify the new data name. Save the data currently being edited to a new data file.

4.9 Summary

This chapter presented in details the user guide to operate the CTT-based item analysis via RASCH-GZ. A complete example with 100 test items was used to generate totally 4 output files with detailed illustrations and explanations. These output files can be held as the excellent examples of developing CTT-based item analysis and significance of this study. In the author's opinion, only strict item analysis can ensure the best quality of test items so as to provide necessary conditions (i.e., well-calibrated test items) for the subsequent test equating. And the examples presented here can also be used as handout for one-day workshop run by the Rasch-GZ program developers to focus on CTT-based item analyses.

In the current era of big data and computer technology development, true testing is based on real and objective data, rather than relying on classroom teaching. Although modern test theory like item response theory (IRT) or Rasch model is well accepted by Chinese professionals of language testing, it is mainly working at seminars or in classrooms. The actual situation in terms of application is by no means optimistic. Bond and Yanzi (2016) undertook a simple literature search in the China National Knowledge Infrastructure[4] (CNKI) with "Rasch Model" as the keyword. The number of

Chinese publications with citations of "Rasch Model" across different disciplines between 1985 and 2016 is very small. According to Bond and Yanzi, in contrast to the number of publication in the social science each year in China which is around 1.1 million, the number of researchers using Rasch Model (in language testing) remains like "a drop in the ocean".

In terms of the application of computer technology to language testing, China's current situation is fully in line with the international practice; however, the testing theory and application remain largely at the level of CTT with linear or descriptive statistics in most schools and universities across the country. In this sense, both CTT and Rasch measurement need further dissemination and promotion.

Notes

1. Rasch-GZ is certified with 2025R11s0879331 by China Copy Right Protection Center.
2. For any detailed information, interested readers may refer to http://www.rasch-gz.com and can also download the short video to get more intuitive yet simple guidance.
3. Due to the limited space and avoidance of repetition, only the first two records are interpreted. For the complete "-ia1" report of the item analysis for 100 test items, interested readers may refer to the appendix at the end of this chapter.
4. http://www.cnki.net

Bibliography

Bachman, L.F., & Adrian, P. (2010). Language Assessment in Practice: Developing Language Assessments and Justifying Their Use in the Real World. Oxford: Oxford University Press.

Bond, T.G., & Fox, C.M. (2015). Applying the Rasch Model: Fundamental Measurement in the Human Science (3rd Ed). New York: Routledge Taylor & Francis Group.

Bond, T.G., Yan, Z., & Heene, M. (2021). Applying the Rasch Model: Fundamental Measurement in the Human Science (4th Ed). New York: Routledge Taylor & Francis Group.

Gui, S.C., Li, W., & Zhang, Q. (1993). The Application of IRT to MET Equating. In NEEA (Ed.), The 4th Annual Forum on Educational Testing in Beijing (pp. 391–393). Beijing, China: China Peace Press.

Hambleton, R.K., Swaminathan, H., & Rogers, H.J. (1985). Item Response Theory: Principle and Application. Academic Publisher.

Kolen, M.J., & Brennan, R.L. (1995). Test Equating: Methods and Practices. New York, USA: Springer-Verlag.

Kolen, M.J., & Brennan, R.L. (2004). Test equating, Scaling, and Linking: Methods and Practices. (2nd Ed). Dordrecht, The Netherlands: Springer-Verlag. GX

Liu, Y. (2020). Test Equating, Scaling, and Linking: Methods and Practices (Kolen, M.J. & Brennan, R.L., Trans.) (3rd Ed). (Original work published 2004)
Linacre, J.M. (2016). A user's Guide to WINSTEPS Rasch-model Computer Programs: Program Manual 3.92.0. ISBN 0-941938-03-4.
Mitchell, R. (2006). "A Guide to Standardized Testing: The Nature of Assessment," centerforpubliceducation.org, Feb. 15, 2006.
Popham, W.J. (1978). Criterion-referenced Measurement. Englewood Cliffs, NJ: Prentice Hall.
Popham, W.J. (1990). Modern Educational Measurement (2nd Ed). Englewood Cliffs, NJ: Prentice Hall
Popham, W.J. (1994). The instructional consequences of criterion-referenced clarity. Educational Measurement: Issues and Practice, 13, 15–18, 30.
Popham, W.J. (1997). What's wrong- and what's right with rubrics. Educational Leadership, 55(2). Retrieved online, May 2002: http://www.ascd.org.readingroom/edlead/9710/popham.html
Popham, W.J. (2005). "Standardized Testing Fails the Exam," edutopia.org, March 23, 2005.
Steiner, S. (2021). "Better measurement and fewer parameters! The true value of Rasch over IRT", The keynote speech at PROMS2021, Nanjing, China.
Thorndike, R.L. (1971). Educational Measurement (2nd Ed). Washington, D.C.: American Council on Education.
Zhang, Q. (2004). Item Analysis and Test Equating: Research and Application. Beijing, China: Higher Education Press.
Zhang, Q. (2012). Towards International Practice of Language Testing in China. Keynote speech given at the PROMS2012, Jiaxing, China, August 6–9, 2012.
Zhang, Q. (2019). Rasch Model: Research and Practice in China. In Myint Swe Khine (Ed.), International Trends in Educational Assessment. Brill | Sense. Retrieved from http://catalog.loc.gov.
Zhang, Zhigan (Ed.) (2021). Biography of Outstanding Social Scientists in Guangdong. Sun-Yat-sen University Press, p. 270. ISBN 978-7-306-07100-2

Appendix III

A Complete Output File of RGZ001.ia1 for 100 Test Items

ITEM ANALYSIS

TEST CODE: basal

TEST DATE: 2025-06-03

The following test information was provided by Prof. Quan Zhang, PhD.

If you have any questions, please contact:

(1) info@rasch-gz.com, or
(2) qzhang141@rasch-gz.com

Test paper part: LST

TEST CODE	ITEM NO.	Pt	Pi	Pd	P	MA	MB	MC	MD	MO
Basal	LST 1	15.46	15.16	10.66	0.72	11.14	10.98	13.59	12.62	–2.48
NO. OF CAN.	KEY	Ar	Br	Cr	Dr	A	B	C	D	O
100	C	–0.25	–0.30	0.31	–0.05	8	12	72	8	0

TEST CODE	ITEM NO.	Pt	Pi	Pd	P	MA	MB	MC	MD	MO
Basal	LST 2	15.46	15.16	16.66	0.18	12.09	12.74	12.81	15.25	–2.48
NO. OF CAN.	KEY	Ar	Br	Cr	Dr	A	B	C	D	O
100	D	–0.20	–0.06	–0.03	0.38	31	34	17	18	0

TEST CODE	ITEM NO.	Pt	Pi	Pd	P	MA	MB	MC	MD	MO
Basal	LST 3	15.46	15.16	14.99	0.31	13.91	13.03	12.98	11.09	–2.48
NO. OF CAN.	KEY	Ar	Br	Cr	Dr	A	B	C	D	O
100	A	0.20	0.01	–0.00	–0.31	31	32	22	15	0

TEST CODE	ITEM NO.	Pt	Pi	Pd	P	MA	MB	MC	MD	MO
Basal	LST 4	15.46	15.16	14.76	0.33	12.74	14.59	11.85	11.38	–2.48
NO. OF CAN.	KEY	Ar	Br	Cr	Dr	A	B	C	D	O
100	B	–0.06	0.36	–0.17	–0.28	37	33	12	18	0

TEST CODE	ITEM NO.	Pt	Pi	Pd	P	MA	MB	MC	MD	MO
Basal	LST 5	15.46	15.16	13.81	0.42	13.28	12.68	13.74	11.15	–2.48
NO. OF CAN.	KEY	Ar	Br	Cr	Dr	A	B	C	D	O
100	C	0.04	–0.07	0.20	–0.28	12	33	42	13	0

TEST CODE	ITEM NO.	Pt	Pi	Pd	P	MA	MB	MC	MD	MO
Basal	LST 6	15.46	15.16	16.97	0.16	12.37	11.50	13.28	14.34	–2.48
NO. OF CAN.	KEY	Ar	Br	Cr	Dr	A	B	C	D	O
100	A	–0.10	–0.25	0.09	0.22	16	17	51	16	0

TEST CODE	ITEM NO.	Pt	Pi	Pd	P	MA	MB	MC	MD	MO
Basal	LST 7	15.46	15.16	16.66	0.18	11.63	14.81	12.64	12.97	–2.48
NO. OF CAN.	KEY	Ar	Br	Cr	Dr	A	B	C	D	O
100	B	–0.22	0.31	–0.07	–0.01	16	18	27	39	0

TEST CODE	ITEM NO.	Pt	Pi	Pd	P	MA	MB	MC	MD	MO
Basal	LST 8	15.46	15.16	16.97	0.16	13.93	12.21	13.28	13.79	−2.48
NO. OF CAN.	KEY	Ar	Br	Cr	Dr	A	B	C	D	O
100	A	0.15	−0.23	0.04	0.15	16	47	14	23	0

TEST CODE	ITEM NO.	Pt	Pi	Pd	P	MA	MB	MC	MD	MO
Basal	LST 9	15.46	15.16	15.71	0.25	12.45	12.10	12.40	14.96	−2.48
NO. OF CAN.	KEY	Ar	Br	Cr	Dr	A	B	C	D	O
100	D	−0.14	−0.16	−0.10	0.39	38	19	18	25	0

TEST CODE	ITEM NO.	Pt	Pi	Pd	P	MA	MB	MC	MD	MO
Basal	LST 10	15.46	15.16	15.46	0.27	12.24	11.51	13.00	15.32	−2.48
NO. OF CAN.	KEY	Ar	Br	Cr	Dr	A	B	C	D	O
100	D	−0.12	−0.35	0.00	0.47	14	35	24	27	0

TEST CODE	ITEM NO.	Pt	Pi	Pd	P	MA	MB	MC	MD	MO
Basal	LST 11	15.46	15.16	14.76	0.33	13.02	13.07	12.04	13.43	−2.48
NO. OF CAN.	KEY	Ar	Br	Cr	Dr	A	B	C	D	O
100	D	0.00	0.01	−0.16	0.10	31	19	17	33	0

TEST CODE	ITEM NO.	Pt	Pi	Pd	P	MA	MB	MC	MD	MO
Basal	LST 12	15.46	15.16	14.65	0.34	11.82	12.25	15.09	11.01	−2.48
NO. OF CAN.	KEY	Ar	Br	Cr	Dr	A	B	C	D	O
100	C	−0.21	−0.18	0.49	−0.29	19	36	34	11	0

TEST CODE	ITEM NO.	Pt	Pi	Pd	P	MA	MB	MC	MD	MO
Basal	LST 13	15.46	15.16	16.66	0.18	13.28	13.82	12.89	12.06	−2.48
NO. OF CAN.	KEY	Ar	Br	Cr	Dr	A	B	C	D	O
100	A	0.05	0.14	−0.04	−0.15	18	17	51	14	0

TEST CODE	ITEM NO.	Pt	Pi	Pd	P	MA	MB	MC	MD	MO
Basal	LST 14	15.46	15.16	14.76	0.33	13.91	13.16	12.68	11.47	−2.48
NO. OF CAN.	KEY	Ar	Br	Cr	Dr	A	B	C	D	O
100	B	0.19	0.04	−0.06	−0.25	29	33	22	16	0

TEST CODE	ITEM NO.	Pt	Pi	Pd	P	MA	MB	MC	MD	MO
Basal	LST 15	15.46	15.16	14.65	0.34	12.36	13.15	13.78	11.79	−2.48
NO. OF CAN.	KEY	Ar	Br	Cr	Dr	A	B	C	D	O
100	C	−0.11	0.03	0.18	−0.19	20	31	34	15	0

TEST CODE	ITEM NO.	Pt	Pi	Pd	P	MA	MB	MC	MD	MO
Basal	LST 16	15.46	15.16	16.51	0.19	12.48	11.96	13.06	15.21	−2.48
NO. OF CAN.	KEY	Ar	Br	Cr	Dr	A	B	C	D	O
100	D	−0.14	−0.19	0.01	0.39	43	20	18	19	0

TEST CODE	ITEM NO.	Pt	Pi	Pd	P	MA	MB	MC	MD	MO
Basal	LST 17	15.46	15.16	15.22	0.29	11.67	12.74	14.59	12.21	−2.48
NO. OF CAN.	KEY	Ar	Br	Cr	Dr	A	B	C	D	O
100	C	−0.23	−0.06	0.34	−0.13	18	37	29	16	0

TEST CODE	ITEM NO.	Pt	Pi	Pd	P	MA	MB	MC	MD	MO
Basal	LST 18	15.46	15.16	16.97	0.16	13.03	11.50	13.36	13.52	−2.48
NO. OF CAN.	KEY	Ar	Br	Cr	Dr	A	B	C	D	O
100	D	0.00	−0.25	0.10	0.09	21	17	46	16	0

TEST CODE	ITEM NO.	Pt	Pi	Pd	P	MA	MB	MC	MD	MO
Basal	LST 19	15.46	15.16	15.58	0.26	13.31	11.96	14.03	11.58	−2.48
NO. OF CAN.	KEY	Ar	Br	Cr	Dr	A	B	C	D	O
100	C	0.08	−0.17	0.21	−0.24	42	15	26	17	0

TEST CODE	ITEM NO.	Pt	Pi	Pd	P	MA	MB	MC	MD	MO
Basal	LST 20	15.46	15.16	16.66	0.18	10.74	11.81	13.38	14.81	−2.48
NO. OF CAN.	KEY	Ar	Br	Cr	Dr	A	B	C	D	O
100	D	−0.35	−0.20	0.12	0.31	14	17	51	18	0

TEST CODE	ITEM NO.	Pt	Pi	Pd	P	MA	MB	MC	MD	MO
Basal	LST 21	15.46	15.16	16.81	0.17	13.28	13.78	12.16	13.04	−2.48
NO. OF CAN.	KEY	Ar	Br	Cr	Dr	A	B	C	D	O
100	A	0.05	0.12	−0.15	0.01	17	13	20	50	0

TEST CODE	ITEM NO.	Pt	Pi	Pd	P	MA	MB	MC	MD	MO
Basal	LST 22	15.46	15.16	15.58	0.26	11.88	13.79	12.22	14.34	–2.48
NO. OF CAN.	KEY	Ar	Br	Cr	Dr	A	B	C	D	O
100	D	–0.18	0.13	–0.21	0.27	15	18	41	26	0

TEST CODE	ITEM NO.	Pt	Pi	Pd	P	MA	MB	MC	MD	MO
Basal	LST 23	15.46	15.16	14.76	0.33	12.19	11.41	14.95	12.62	–2.48
NO. OF CAN.	KEY	Ar	Br	Cr	Dr	A	B	C	D	O
100	C	–0.14	–0.32	0.44	–0.07	17	26	33	24	0

TEST CODE	ITEM NO.	Pt	Pi	Pd	P	MA	MB	MC	MD	MO
Basal	LST 24	15.46	15.16	14.22	0.38	13.14	11.09	11.78	14.07	16.56
NO. OF CAN.	KEY	Ar	Br	Cr	Dr	A	B	C	D	O
100	D	0.03	–0.33	–0.19	0.27	28	18	14	38	2

TEST CODE	ITEM NO.	Pt	Pi	Pd	P	MA	MB	MC	MD	MO
Basal	LST 25	15.46	15.16	15.96	0.23	12.64	11.96	14.19	13.12	–2.48
NO. OF CAN.	KEY	Ar	Br	Cr	Dr	A	B	C	D	O
100	C	–0.09	–0.17	0.23	0.02	35	17	23	25	0

TEST CODE	ITEM NO.	Pt	Pi	Pd	P	MA	MB	MC	MD	MO
Basal	LST 26	15.46	15.16	17.13	0.15	13.23	13.63	10.82	13.64	–2.48
NO. OF CAN.	KEY	Ar	Br	Cr	Dr	A	B	C	D	O
100	B	0.07	0.10	–0.35	0.11	52	15	15	18	0

TEST CODE	ITEM NO.	Pt	Pi	Pd	P	MA	MB	MC	MD	MO
Basal	LST 27	15.46	15.16	15.71	0.25	11.37	14.17	13.57	12.81	–2.48
NO. OF CAN.	KEY	Ar	Br	Cr	Dr	A	B	C	D	O
100	B	–0.29	0.23	0.10	–0.05	20	25	18	37	0

TEST CODE	ITEM NO.	Pt	Pi	Pd	P	MA	MB	MC	MD	MO
Basal	LST 28	15.46	15.16	15.10	0.30	14.46	12.72	11.66	13.52	–2.48
NO. OF CAN.	KEY	Ar	Br	Cr	Dr	A	B	C	D	O
100	A	0.31	–0.05	–0.32	0.09	30	19	35	16	0

TEST CODE	ITEM NO.	Pt	Pi	Pd	P	MA	MB	MC	MD	MO
Basal	LST 29	15.46	15.16	15.34	0.28	14.73	12.70	11.52	12.58	−2.48
NO. OF CAN.	KEY	Ar	Br	Cr	Dr	A	B	C	D	O
100	A	0.36	−0.07	−0.27	−0.07	28	34	21	17	0

TEST CODE	ITEM NO.	Pt	Pi	Pd	P	MA	MB	MC	MD	MO
Basal	LST 30	15.46	15.16	15.34	0.28	14.17	12.37	13.59	10.77	−2.48
NO. OF CAN.	KEY	Ar	Br	Cr	Dr	A	B	C	D	O
100	A	0.24	−0.10	0.15	−0.41	28	13	38	21	0

TEST CODE	ITEM NO.	Pt	Pi	Pd	P	MA	MB	MC	MD	MO
Basal	LST 31	15.46	15.16	14.22	0.38	14.73	11.40	13.07	11.49	−2.48
NO. OF CAN.	KEY	Ar	Br	Cr	Dr	A	B	C	D	O
100	A	0.43	−0.29	0.01	−0.28	38	21	19	22	0

TEST CODE	ITEM NO.	Pt	Pi	Pd	P	MA	MB	MC	MD	MO
Basal	LST 32	15.46	15.16	12.90	0.51	13.20	12.09	12.33	13.53	−2.48
NO. OF CAN.	KEY	Ar	Br	Cr	Dr	A	B	C	D	O
100	A	0.06	−0.13	−0.12	0.10	51	10	18	21	0

TEST CODE	ITEM NO.	Pt	Pi	Pd	P	MA	MB	MC	MD	MO
Basal	LST 33	15.46	15.16	15.58	0.26	13.78	12.67	13.48	11.45	−2.48
NO. OF CAN.	KEY	Ar	Br	Cr	Dr	A	B	C	D	O
100	A	0.16	−0.06	0.11	−0.27	26	24	32	18	0

TEST CODE	ITEM NO.	Pt	Pi	Pd	P	MA	MB	MC	MD	MO
Basal	LST 34	15.46	15.16	4.17	1.00	−2.48	−2.48	−2.48	13.00	−2.48
NO. OF CAN.	KEY	Ar	Br	Cr	Dr	A	B	C	D	O
100	D	0.00	0.00	0.00	0.00	0	0	0	100	0

TEST CODE	ITEM NO.	Pt	Pi	Pd	P	MA	MB	MC	MD	MO
Basal	LST 35	15.46	15.16	15.83	0.24	14.00	12.28	11.96	14.21	−2.48
NO. OF CAN.	KEY	Ar	Br	Cr	Dr	A	B	C	D	O
100	D	0.18	−0.15	−0.21	0.23	20	29	27	24	0

TEST CODE	ITEM NO.	Pt	Pi	Pd	P	MA	MB	MC	MD	MO
Basal	LST 36	15.46	15.16	16.09	0.22	13.21	12.28	12.76	14.05	−2.48
NO. OF CAN.	KEY	Ar	Br	Cr	Dr	A	B	C	D	O
100	D	0.04	−0.15	−0.05	0.20	21	29	28	22	0

TEST CODE	ITEM NO.	Pt	Pi	Pd	P	MA	MB	MC	MD	MO
Basal	LST 37	15.46	15.16	15.46	0.27	13.71	12.40	12.74	13.10	−2.48
NO. OF CAN.	KEY	Ar	Br	Cr	Dr	A	B	C	D	O
100	A	0.15	−0.12	−0.05	0.02	27	24	27	22	0

TEST CODE	ITEM NO.	Pt	Pi	Pd	P	MA	MB	MC	MD	MO
Basal	LST 38	15.46	15.16	15.83	0.24	13.62	11.87	12.42	14.53	−2.48
NO. OF CAN.	KEY	Ar	Br	Cr	Dr	A	B	C	D	O
100	D	0.11	−0.24	−0.12	0.30	19	28	29	24	0

TEST CODE	ITEM NO.	Pt	Pi	Pd	P	MA	MB	MC	MD	MO
Basal	LST 39	15.46	15.16	16.23	0.21	12.77	12.75	15.40	11.39	−2.48
NO. OF CAN.	KEY	Ar	Br	Cr	Dr	A	B	C	D	O
100	C	−0.05	−0.05	0.44	−0.30	31	25	21	23	0

TEST CODE	ITEM NO.	Pt	Pi	Pd	P	MA	MB	MC	MD	MO
Basal	LST 40	15.46	15.16	15.71	0.25	12.80	12.52	13.93	13.01	−2.48
NO. OF CAN.	KEY	Ar	Br	Cr	Dr	A	B	C	D	O
100	A	−0.04	−0.11	0.17	0.00	25	33	22	20	0

Test paper part: GRM

TEST CODE	ITEM NO.	Pt	Pi	Pd	P	MA	MB	MC	MD	MO
Basal	GRM 41	15.46	15.64	15.96	0.23	13.21	14.59	12.57	11.31	−0.94
NO. OF CAN.	KEY	Ar	Br	Cr	Dr	A	B	C	D	O
100	B	0.05	0.30	−0.09	−0.29	30	23	29	18	0

TEST CODE	ITEM NO.	Pt	Pi	Pd	P	MA	MB	MC	MD	MO
Basal	GRM 42	15.46	15.64	14.65	0.34	14.01	13.20	11.90	12.37	−0.94
NO. OF CAN.	KEY	Ar	Br	Cr	Dr	A	B	C	D	O
100	A	0.24	0.04	−0.20	−0.12	34	21	22	23	0

TEST CODE	ITEM NO.	Pt	Pi	Pd	P	MA	MB	MC	MD	MO
Basal	GRM 43	15.46	15.64	15.58	0.26	12.02	13.32	14.06	12.29	–0.94
NO. OF CAN.	KEY	Ar	Br	Cr	Dr	A	B	C	D	O
100	C	–0.17	0.07	0.21	–0.15	18	29	26	27	0

TEST CODE	ITEM NO.	Pt	Pi	Pd	P	MA	MB	MC	MD	MO
Basal	GRM 44	15.46	15.64	16.09	0.22	12.12	11.46	15.13	13.89	–0.94
NO. OF CAN.	KEY	Ar	Br	Cr	Dr	A	B	C	D	O
100	C	–0.19	–0.31	0.40	0.16	30	26	22	22	0

TEST CODE	ITEM NO.	Pt	Pi	Pd	P	MA	MB	MC	MD	MO
Basal	GRM 45	15.46	15.64	16.09	0.22	13.26	12.59	11.73	14.55	–0.94
NO. OF CAN.	KEY	Ar	Br	Cr	Dr	A	B	C	D	O
100	D	0.05	–0.09	–0.23	0.29	24	33	21	22	0

TEST CODE	ITEM NO.	Pt	Pi	Pd	P	MA	MB	MC	MD	MO
Basal	GRM 46	15.46	15.64	16.51	0.19	13.33	14.13	11.99	12.68	–0.94
NO. OF CAN.	KEY	Ar	Br	Cr	Dr	A	B	C	D	O
100	D	0.06	0.23	–0.22	–0.06	24	26	31	19	0

TEST CODE	ITEM NO.	Pt	Pi	Pd	P	MA	MB	MC	MD	MO
Basal	GRM 47	15.46	15.64	16.09	0.22	13.12	12.67	13.89	12.48	–0.94
NO. OF CAN.	KEY	Ar	Br	Cr	Dr	A	B	C	D	O
100	C	0.02	–0.05	0.16	–0.12	28	17	22	33	0

TEST CODE	ITEM NO.	Pt	Pi	Pd	P	MA	MB	MC	MD	MO
Basal	GRM 48	15.46	15.64	16.23	0.21	13.81	12.82	12.38	12.94	–0.94
NO. OF CAN.	KEY	Ar	Br	Cr	Dr	A	B	C	D	O
100	A	0.15	–0.04	–0.10	–0.01	21	29	16	34	0

TEST CODE	ITEM NO.	Pt	Pi	Pd	P	MA	MB	MC	MD	MO
Basal	GRM 49	15.46	15.64	15.58	0.26	13.50	13.29	13.39	11.89	–0.94
NO. OF CAN.	KEY	Ar	Br	Cr	Dr	A	B	C	D	O
100	B	0.10	0.06	0.07	–0.22	26	26	22	26	0

TEST CODE	ITEM NO.	Pt	Pi	Pd	P	MA	MB	MC	MD	MO
Basal	GRM 50	15.46	15.64	15.34	0.28	12.69	12.04	11.11	15.92	−0.94
NO. OF CAN.	KEY	Ar	Br	Cr	Dr	A	B	C	D	O
100	D	−0.06	−0.16	−0.42	0.61	25	16	31	28	0

TEST CODE	ITEM NO.	Pt	Pi	Pd	P	MA	MB	MC	MD	MO
Basal	GRM 51	15.46	15.64	15.96	0.23	12.08	14.43	13.31	12.25	−0.94
NO. OF CAN.	KEY	Ar	Br	Cr	Dr	A	B	C	D	O
100	B	−0.19	0.27	0.07	−0.14	28	23	28	21	0

TEST CODE	ITEM NO.	Pt	Pi	Pd	P	MA	MB	MC	MD	MO
Basal	GRM 52	15.46	15.64	15.46	0.27	13.76	13.08	13.37	11.93	−0.94
NO. OF CAN.	KEY	Ar	Br	Cr	Dr	A	B	C	D	O
100	C	0.16	0.01	0.08	−0.23	29	13	27	31	0

TEST CODE	ITEM NO.	Pt	Pi	Pd	P	MA	MB	MC	MD	MO
Basal	GRM 53	15.46	15.64	15.22	0.29	11.71	15.21	13.06	11.64	−0.94
NO. OF CAN.	KEY	Ar	Br	Cr	Dr	A	B	C	D	O
100	B	−0.22	0.47	0.01	−0.30	18	29	22	31	0

TEST CODE	ITEM NO.	Pt	Pi	Pd	P	MA	MB	MC	MD	MO
Basal	GRM 54	15.46	15.64	14.99	0.31	14.58	12.56	12.19	12.02	−0.94
NO. OF CAN.	KEY	Ar	Br	Cr	Dr	A	B	C	D	O
100	A	0.35	−0.09	−0.16	−0.17	31	27	24	18	0

TEST CODE	ITEM NO.	Pt	Pi	Pd	P	MA	MB	MC	MD	MO
Basal	GRM 55	15.46	15.64	15.58	0.26	12.42	12.69	14.20	12.64	−0.94
NO. OF CAN.	KEY	Ar	Br	Cr	Dr	A	B	C	D	O
100	C	−0.12	−0.06	0.24	−0.07	27	25	26	22	0

TEST CODE	ITEM NO.	Pt	Pi	Pd	P	MA	MB	MC	MD	MO
Basal	GRM 56	15.46	15.64	15.71	0.25	13.22	13.57	12.27	12.82	−0.94
NO. OF CAN.	KEY	Ar	Br	Cr	Dr	A	B	C	D	O
100	B	0.04	0.11	−0.13	−0.04	26	25	20	29	0

TEST CODE	ITEM NO.	Pt	Pi	Pd	P	MA	MB	MC	MD	MO
Basal	GRM 57	15.46	15.64	15.96	0.23	12.62	14.75	12.09	12.97	−0.94
NO. OF CAN.	KEY	Ar	Br	Cr	Dr	A	B	C	D	O
100	B	−0.08	0.33	−0.21	−0.01	25	23	33	19	0

TEST CODE	ITEM NO.	Pt	Pi	Pd	P	MA	MB	MC	MD	MO
Basal	GRM 58	15.46	15.64	15.46	0.27	11.58	13.37	13.08	13.87	−0.94
NO. OF CAN.	KEY	Ar	Br	Cr	Dr	A	B	C	D	O
100	B	−0.27	0.08	0.02	0.17	23	27	26	24	0

TEST CODE	ITEM NO.	Pt	Pi	Pd	P	MA	MB	MC	MD	MO
Basal	GRM 59	15.46	15.64	16.51	0.19	14.31	11.36	13.27	13.22	−0.94
NO. OF CAN.	KEY	Ar	Br	Cr	Dr	A	B	C	D	O
100	A	0.23	−0.32	0.06	0.04	19	24	35	22	0

TEST CODE	ITEM NO.	Pt	Pi	Pd	P	MA	MB	MC	MD	MO
Basal	GRM 60	15.46	15.64	15.96	0.23	13.89	12.48	13.72	11.82	−0.94
NO. OF CAN.	KEY	Ar	Br	Cr	Dr	A	B	C	D	O
100	C	0.19	−0.10	0.14	−0.24	29	22	23	26	0

TEST CODE	ITEM NO.	Pt	Pi	Pd	P	MA	MB	MC	MD	MO
Basal	GRM 61	15.46	15.64	15.10	0.30	11.91	14.17	13.27	12.81	−0.94
NO. OF CAN.	KEY	Ar	Br	Cr	Dr	A	B	C	D	O
100	C	−0.20	0.20	0.06	−0.04	20	17	30	33	0

TEST CODE	ITEM NO.	Pt	Pi	Pd	P	MA	MB	MC	MD	MO
Basal	GRM 62	15.46	15.64	15.83	0.24	12.69	13.79	12.02	13.64	−0.94
NO. OF CAN.	KEY	Ar	Br	Cr	Dr	A	B	C	D	O
100	B	−0.06	0.15	−0.20	0.12	25	24	27	24	0

TEST CODE	ITEM NO.	Pt	Pi	Pd	P	MA	MB	MC	MD	MO
Basal	GRM 63	15.46	15.64	15.22	0.29	11.18	15.34	12.03	12.73	−0.94
NO. OF CAN.	KEY	Ar	Br	Cr	Dr	A	B	C	D	O
100	B	−0.33	0.50	−0.19	−0.05	20	29	25	26	0

TEST CODE	ITEM NO.	Pt	Pi	Pd	P	MA	MB	MC	MD	MO
Basal	GRM 64	15.46	15.64	15.34	0.28	14.33	12.73	11.48	13.70	−0.94
NO. OF CAN.	KEY	Ar	Br	Cr	Dr	A	B	C	D	O
100	D	0.24	−0.05	−0.31	0.15	21	24	27	28	0

TEST CODE	ITEM NO.	Pt	Pi	Pd	P	MA	MB	MC	MD	MO
Basal	GRM 65	15.46	15.64	16.23	0.21	13.72	13.77	13.27	11.32	−0.94
NO. OF CAN.	KEY	Ar	Br	Cr	Dr	A	B	C	D	O
100	A	0.13	0.16	0.05	−0.34	21	28	25	26	0

TEST CODE	ITEM NO.	Pt	Pi	Pd	P	MA	MB	MC	MD	MO
Basal	GRM 66	15.46	15.64	14.43	0.36	12.34	13.18	13.89	12.18	−0.94
NO. OF CAN.	KEY	Ar	Br	Cr	Dr	A	B	C	D	O
100	C	−0.14	0.03	0.21	−0.15	28	16	36	20	0

TEST CODE	ITEM NO.	Pt	Pi	Pd	P	MA	MB	MC	MD	MO
Basal	GRM 67	15.46	15.64	15.46	0.27	12.13	14.04	12.60	13.14	−0.94
NO. OF CAN.	KEY	Ar	Br	Cr	Dr	A	B	C	D	O
100	B	−0.16	0.21	−0.08	0.03	23	27	28	22	0

TEST CODE	ITEM NO.	Pt	Pi	Pd	P	MA	MB	MC	MD	MO
Basal	GRM 68	15.46	15.64	15.96	0.23	12.34	14.19	11.82	13.21	−0.94
NO. OF CAN.	KEY	Ar	Br	Cr	Dr	A	B	C	D	O
100	B	−0.12	0.23	−0.20	0.05	21	23	18	38	0

TEST CODE	ITEM NO.	Pt	Pi	Pd	P	MA	MB	MC	MD	MO
Basal	GRM 69	15.46	15.64	15.10	0.30	12.54	12.49	13.88	12.80	−0.94
NO. OF CAN.	KEY	Ar	Br	Cr	Dr	A	B	C	D	O
100	C	−0.09	−0.09	0.19	−0.04	25	19	30	26	0

TEST CODE	ITEM NO.	Pt	Pi	Pd	P	MA	MB	MC	MD	MO
Basal	GRM 70	15.46	15.64	16.23	0.21	12.56	14.59	12.03	13.40	−0.94
NO. OF CAN.	KEY	Ar	Br	Cr	Dr	A	B	C	D	O
100	B	−0.08	0.29	−0.23	0.08	22	21	34	23	0

Test paper part: RDG

TEST CODE	ITEM NO.	Pt	Pi	Pd	P	MA	MB	MC	MD	MO
Basal	RDG 71	15.46	15.68	15.71	0.25	12.43	14.83	13.33	11.38	0.50
NO. OF CAN.	KEY	Ar	Br	Cr	Dr	A	B	C	D	O
100	B	–0.12	0.36	0.06	–0.30	31	25	22	22	0

TEST CODE	ITEM NO.	Pt	Pi	Pd	P	MA	MB	MC	MD	MO
Basal	RDG 72	15.46	15.68	15.83	0.24	13.99	13.18	13.64	11.16	0.50
NO. OF CAN.	KEY	Ar	Br	Cr	Dr	A	B	C	D	O
100	C	0.19	0.04	0.12	–0.35	24	28	24	24	0

TEST CODE	ITEM NO.	Pt	Pi	Pd	P	MA	MB	MC	MD	MO
Basal	RDG 73	15.46	15.68	14.87	0.32	13.60	13.87	12.23	11.36	0.50
NO. OF CAN.	KEY	Ar	Br	Cr	Dr	A	B	C	D	O
100	B	0.13	0.19	–0.13	–0.29	30	32	19	19	0

TEST CODE	ITEM NO.	Pt	Pi	Pd	P	MA	MB	MC	MD	MO
Basal	RDG 74	15.46	15.68	15.34	0.28	13.88	14.12	11.70	12.69	0.50
NO. OF CAN.	KEY	Ar	Br	Cr	Dr	A	B	C	D	O
100	B	0.15	0.23	–0.29	–0.06	19	28	32	21	0

TEST CODE	ITEM NO.	Pt	Pi	Pd	P	MA	MB	MC	MD	MO
Basal	RDG 75	15.46	15.68	15.83	0.24	10.80	14.33	13.08	13.46	0.50
NO. OF CAN.	KEY	Ar	Br	Cr	Dr	A	B	C	D	O
100	B	–0.40	0.26	0.02	0.09	21	24	29	26	0

TEST CODE	ITEM NO.	Pt	Pi	Pd	P	MA	MB	MC	MD	MO
Basal	RDG 76	15.46	15.68	16.37	0.20	12.72	13.05	13.23	13.16	0.50
NO. OF CAN.	KEY	Ar	Br	Cr	Dr	A	B	C	D	O
100	B	–0.07	0.01	0.04	0.03	35	20	24	21	0

TEST CODE	ITEM NO.	Pt	Pi	Pd	P	MA	MB	MC	MD	MO
Basal	RDG 77	15.46	15.68	16.09	0.22	12.94	13.49	12.50	13.18	0.50
NO. OF CAN.	KEY	Ar	Br	Cr	Dr	A	B	C	D	O
100	B	–0.01	0.09	–0.10	0.03	30	22	26	22	0

TEST CODE	ITEM NO.	Pt	Pi	Pd	P	MA	MB	MC	MD	MO
Basal	RDG 78	15.46	15.68	16.09	0.22	11.92	14.26	14.46	12.35	0.50
NO. OF CAN.	KEY	Ar	Br	Cr	Dr	A	B	C	D	O
100	C	−0.26	0.22	0.27	−0.12	37	18	22	23	0

TEST CODE	ITEM NO.	Pt	Pi	Pd	P	MA	MB	MC	MD	MO
Basal	RDG 79	15.46	15.68	15.83	0.24	12.32	12.36	14.81	12.61	0.50
NO. OF CAN.	KEY	Ar	Br	Cr	Dr	A	B	C	D	O
100	C	−0.13	−0.13	0.35	−0.08	25	27	24	24	0

TEST CODE	ITEM NO.	Pt	Pi	Pd	P	MA	MB	MC	MD	MO
Basal	RDG 80	15.46	15.68	16.09	0.22	13.33	11.87	12.97	14.09	0.50
NO. OF CAN.	KEY	Ar	Br	Cr	Dr	A	B	C	D	O
100	D	0.06	−0.23	−0.01	0.20	22	27	29	22	0

TEST CODE	ITEM NO.	Pt	Pi	Pd	P	MA	MB	MC	MD	MO
Basal	RDG 81	15.46	15.68	15.83	0.24	12.44	14.33	12.85	12.45	0.50
NO. OF CAN.	KEY	Ar	Br	Cr	Dr	A	B	C	D	O
100	B	−0.12	0.26	−0.03	−0.10	30	24	25	21	0

TEST CODE	ITEM NO.	Pt	Pi	Pd	P	MA	MB	MC	MD	MO
Basal	RDG 82	15.46	15.68	15.34	0.28	11.55	12.42	14.71	13.14	0.50
NO. OF CAN.	KEY	Ar	Br	Cr	Dr	A	B	C	D	O
100	C	−0.29	−0.11	0.36	0.03	26	23	28	23	0

TEST CODE	ITEM NO.	Pt	Pi	Pd	P	MA	MB	MC	MD	MO
Basal	RDG 83	15.46	15.68	15.58	0.26	11.94	13.01	12.00	15.17	0.50
NO. OF CAN.	KEY	Ar	Br	Cr	Dr	A	B	C	D	O
100	D	−0.22	0.00	−0.21	0.44	27	19	28	26	0

TEST CODE	ITEM NO.	Pt	Pi	Pd	P	MA	MB	MC	MD	MO
Basal	RDG 84	15.46	15.68	16.23	0.21	13.34	14.81	12.72	11.47	0.50
NO. OF CAN.	KEY	Ar	Br	Cr	Dr	A	B	C	D	O
100	B	0.07	0.33	−0.05	−0.32	31	21	20	28	0

TEST CODE	ITEM NO.	Pt	Pi	Pd	P	MA	MB	MC	MD	MO
Basal	RDG 85	15.46	15.68	16.81	0.17	14.39	12.47	12.98	12.57	0.50
NO. OF CAN.	KEY	Ar	Br	Cr	Dr	A	B	C	D	O
100	A	0.23	–0.09	–0.00	–0.09	17	20	34	29	0

TEST CODE	ITEM NO.	Pt	Pi	Pd	P	MA	MB	MC	MD	MO
Basal	RDG 86	15.46	15.68	16.97	0.16	12.83	10.15	13.30	14.43	0.50
NO. OF CAN.	KEY	Ar	Br	Cr	Dr	A	B	C	D	O
100	D	–0.03	–0.44	0.11	0.23	15	13	56	16	0

TEST CODE	ITEM NO.	Pt	Pi	Pd	P	MA	MB	MC	MD	MO
Basal	RDG 87	15.46	15.68	16.97	0.16	13.09	10.79	15.16	12.43	0.50
NO. OF CAN.	KEY	Ar	Br	Cr	Dr	A	B	C	D	O
100	C	0.03	–0.34	0.35	–0.10	53	13	16	18	0

TEST CODE	ITEM NO.	Pt	Pi	Pd	P	MA	MB	MC	MD	MO
Basal	RDG 88	15.46	15.68	17.69	0.12	11.87	12.96	12.71	15.50	0.50
NO. OF CAN.	KEY	Ar	Br	Cr	Dr	A	B	C	D	O
100	D	–0.16	–0.01	–0.14	0.37	9	11	68	12	0

TEST CODE	ITEM NO.	Pt	Pi	Pd	P	MA	MB	MC	MD	MO
Basal	RDG 89	15.46	15.68	16.51	0.19	14.14	12.37	12.81	12.82	0.50
NO. OF CAN.	KEY	Ar	Br	Cr	Dr	A	B	C	D	O
100	A	0.20	–0.10	–0.03	–0.06	19	16	11	54	0

TEST CODE	ITEM NO.	Pt	Pi	Pd	P	MA	MB	MC	MD	MO
Basal	RDG 90	15.46	15.68	14.43	0.36	11.96	11.56	12.97	13.89	0.50
NO. OF CAN.	KEY	Ar	Br	Cr	Dr	A	B	C	D	O
100	D	–0.17	–0.20	–0.01	0.22	16	10	38	36	0

TEST CODE	ITEM NO.	Pt	Pi	Pd	P	MA	MB	MC	MD	MO
Basal	RDG 91	15.46	15.68	15.46	0.27	9.31	12.21	14.81	14.22	0.50
NO. OF CAN.	KEY	Ar	Br	Cr	Dr	A	B	C	D	O
100	C	–0.59	–0.18	0.37	0.24	15	32	27	26	0

TEST CODE	ITEM NO.	Pt	Pi	Pd	P	MA	MB	MC	MD	MO
Basal	RDG 92	15.46	15.68	14.99	0.31	13.60	11.14	12.51	13.87	0.50
NO. OF CAN.	KEY	Ar	Br	Cr	Dr	A	B	C	D	O
100	D	0.13	−0.32	−0.09	0.19	29	18	22	31	0

TEST CODE	ITEM NO.	Pt	Pi	Pd	P	MA	MB	MC	MD	MO
Basal	RDG 93	15.46	15.68	15.58	0.26	11.89	13.14	13.46	13.88	0.50
NO. OF CAN.	KEY	Ar	Br	Cr	Dr	A	B	C	D	O
100	C	−0.24	0.03	0.09	0.15	29	26	26	19	0

TEST CODE	ITEM NO.	Pt	Pi	Pd	P	MA	MB	MC	MD	MO
Basal	RDG 94	15.46	15.68	17.13	0.15	13.26	14.92	11.23	11.99	0.50
NO. OF CAN.	KEY	Ar	Br	Cr	Dr	A	B	C	D	O
100	B	0.08	0.31	−0.25	−0.20	51	15	10	24	0

TEST CODE	ITEM NO.	Pt	Pi	Pd	P	MA	MB	MC	MD	MO
Basal	RDG 95	15.46	15.68	15.22	0.29	12.88	14.51	11.88	12.45	0.50
NO. OF CAN.	KEY	Ar	Br	Cr	Dr	A	B	C	D	O
100	B	−0.02	0.32	−0.20	−0.14	14	29	19	38	0

TEST CODE	ITEM NO.	Pt	Pi	Pd	P	MA	MB	MC	MD	MO
Basal	RDG 96	15.46	15.68	15.46	0.27	14.44	13.35	11.56	12.80	0.50
NO. OF CAN.	KEY	Ar	Br	Cr	Dr	A	B	C	D	O
100	A	0.30	0.07	−0.31	−0.04	27	23	30	20	0

TEST CODE	ITEM NO.	Pt	Pi	Pd	P	MA	MB	MC	MD	MO
Basal	RDG 97	15.46	15.68	16.09	0.22	15.89	12.39	11.98	11.96	0.50
NO. OF CAN.	KEY	Ar	Br	Cr	Dr	A	B	C	D	O
100	A	0.54	−0.16	−0.19	−0.17	22	40	21	17	0

TEST CODE	ITEM NO.	Pt	Pi	Pd	P	MA	MB	MC	MD	MO
Basal	RDG 98	15.46	15.68	15.34	0.28	13.95	12.19	13.07	11.97	0.50
NO. OF CAN.	KEY	Ar	Br	Cr	Dr	A	B	C	D	O
100	A	0.20	−0.12	0.02	−0.18	28	13	41	18	0

TEST CODE	ITEM NO.	Pt	Pi	Pd	P	MA	MB	MC	MD	MO
Basal	RDG 99	15.46	15.68	13.40	0.46	14.11	11.17	13.50	11.60	0.50
NO. OF CAN.	KEY	Ar	Br	Cr	Dr	A	B	C	D	O
100	A	0.32	−0.28	0.08	−0.28	46	13	16	25	0

TEST CODE	ITEM NO.	Pt	Pi	Pd	P	MA	MB	MC	MD	MO
Basal	RDG 100	15.46	15.68	13.20	0.48	13.33	12.77	13.03	12.35	0.50
NO. OF CAN.	KEY	Ar	Br	Cr	Dr	A	B	C	D	O
100	A	0.10	−0.03	0.01	−0.12	48	7	22	23	0

Rasch-GZ

— End

Appendix IV

The Complete TWCT (P-value and R-bis) of a Test Paper

The output file of RGZ001.ia2 is TWCT, consisting of four parts. Part One is the table showing P-value and R-bis of the whole test paper; Part Two specifies P-value and R-bis of test items in each section of the test paper; Part Three is a table of correlation matrix; and Part Four is a table showing factor loadings.

1. Table of P-value and R-bis. for All the Test Items within a Test Paper

R/P	0–0.1	0.1–0.2	0.2–0.3	0.3–0.4	0.4–0.5	0.5–0.6	0.6–0.7	0.7–0.8	0.8–0.9	0.9–1	Total
<0.1	0	6	7	2	1	1	0	0	0	1	18
0.1–0.2	0	2	13	3	0	0	0	0	0	0	18
0.2–0.3	0	4	20	5	1	0	0	0	0	0	30
0.3–0.4	0	7	13	2	1	0	0	1	0	0	24
0.4–0.5	0	0	5	3	0	0	0	0	0	0	8
0.5–0.6	0	0	1	0	0	0	0	0	0	0	1
0.6–0.7	0	0	1	0	0	0	0	0	0	0	1
0.7–0.8	0	0	0	0	0	0	0	0	0	0	0
0.8–0.9	0	0	0	0	0	0	0	0	0	0	0
0.9–1	0	0	0	0	0	0	0	0	0	0	0
Total:	0	19	60	15	3	1	0	1	0	1	100
	VD (5%)	**D (15%)**		**I (60%)**				**E (15%)**		**VE (5%)**	**Total**
TOTAL	5	15		60				15		5	100
TOTAL	0	79		19				1		1	100
PASS	0	27		6				1		0	34

Mean	**SD**	**Varn.**	**p+**	**pd**	**R11**	**Rbis**	**aVALUE**	**Skew.**	**Kurt.**
27.01	5.29	28.03	0.27	15.46	0.34	0.16	0.44	0.73	0.89

In this Two-way Checklist Table (TWCT), the horizontal line indicates the P-value (0-1); vertical line shows the discrimination index (between 0.1 and 1). In this table, attention must be paid to the three lines each beginning with TOTAL, TOTAL and PASS at the bottom of the table.

Here the first TOTAL refers to the expected number of test items in a normal distribution where very difficult (VD) test items should account for 5%; difficult (D) test items should account for 15%; intermediate (I) test items should account for 60%; easy (E) test items, for 15%, and very easy (VE) test items, 5%.

The second TOTAL refers to the actual numbers of test items at each level against the corresponding numbers required in a normal distribution. Here we've actually obtained 0 items for VD, but 79 items for D, 19 items for I, 1 item for E and 1 item for VE.

The line beginning with PASS refers to the numbers of test items that meet the requirements of both P-value and R-bis within a test paper. Here we could see, no item meets the requirement of VD; only 27 out of 79 items meet the requirements for D; 6 out of 19 items meet the requirements for I; only 1 item meets the requirements for E, and no item meets the requirements for VE. In this way, the TWCT generated by the Rasch-GZ gives a clear picture of whether the test design conforms to the requirements that the test developers desired. Meanwhile, the TWCT also offers specific guideline to test developers regarding how to adjust the overall difficulty level of a test paper. This is the prerequisite condition for follow-up test equating and item banking. Now with the TWCT, test developers could feasibly handle such difficulty adjustments.

2. Table of P-value and R-bis. for Items within Each Section Section Name: LST (40 items)

Mean	**SD**	**Varn.**	**p+**	**pd**	**R11**	**Rbis**	**Skew.**	**Kurt.**
11.79	3.05	9.28	0.29	15.16	0.21	0.22	0.75	0.91
Difficulty	**Total**	**NO. (<0.3)**	**ITEMS**					
VD	0	0						
D	26	17	6, 8, 13, 18, 19, 21, 22, 25, 26, 27, 30, 33, 35, 36, 37, 38, 40					
I	12	7	3, 5, 11, 14, 15, 24, 32					
E	1	0						
VE	1	1	34					

This is the table of P-value and R-bis. for test items in Listening Comprehension Part. As can be seen, no very difficult items. There are totally 26 difficult items, but 17 ones do not meet the requirements, or in other words, the discrimination indexes of these 17 items are lower than 0.3 and are specifically listed; there are totally 12 intermediate items, out of which 7 items do not meet the requirements, and the discrimination indexes of these 7 items are lower than 0.3 and are specifically listed. In this listening part, there is only 1 easy item, and this easy item meets the requirement, or rather, its discrimination index is >=0.3. There is 1 very easy item but this very easy item (specified as Item 34) does not meet the requirement.

Section Name: GRM (30 items)

Mean	SD	Varn.	p+	pd	R11	Rbis	Skew.	Kurt.
7.65	2.19	4.82	0.26	15.64	−0.18	0.23	−0.05	0.32
Difficulty	**Total**	**NO.(<0.3)**	**Items**					
VD	0	0						
D	25	19	43, 45, 46, 47, 48, 49, 51, 52, 55, 56, 58, 59, 60, 62, 64, 65, 67, 68, 70					
I	5	4	42, 61, 66, 69					
E	0	0						
VE	0	0						

This is the table of P-value and R-bis. for test items in grammar part. As shown in the table above, no very difficult (VD) items. There are totally 25 difficult items, but 19 ones do not meet the requirements because their discrimination indexes are lower than 0.3 and are thus listed in order. There are totally 5 intermediate items, out of which 4 items do not meet the requirements, and the discrimination indexes of these 4 items are lower than 0.3 and are listed in order. In this listening part, there are no easy and very easy items.

Section Name: RDG (30 items)

Mean	SD	Varn.	p+	pd	R11	Rbis	Skew.	Kurt.
7.57	2.42	5.86	0.25	15.67	0.07	0.25	−0.14	0.22
Difficulty	**Total**	**NO.(<0.3)**	**Items**					
VD	0	0						
D	25	14	72, 74, 75, 76, 77, 78, 80, 81, 85, 86, 89, 93, 96, 98					
I	5	4	73, 90, 92, 100					
E	0	0						
VE	0	0						

This is the table of P-value and R-bis. for test items in Reading Comprehension Part. As shown in the table above, no very difficult items. There are totally 25 difficult items, but 14 ones out of which do not meet the requirements; the discrimination indexes of these 14 items are lower than 0.3 and are listed in order. There are totally 5 intermediate items, out of which 4 ones do not meet the requirements, and the discrimination indexes of these 4 items are lower than 0.3 and are listed in order: Item 73, Item 90, Item 92, and Item 100. In this part, there are no easy nor very easy items.

3. Correlation Matrix

	0	I	II	III
0	1.00	0.82	0.50	0.70
1		1.00	0.14	0.41
2			1.00	0.02
3				1.00
Mean	27.01	11.79	7.65	7.57
SD	5.29	3.05	2.19	2.42

0 = FACTOR
1 = LST, 2 = GRM, 3 = RDG

4. Factor Loadings

	FACTOR 1	FACTOR 2	FACTOR 3
1	0.690	0.117	0.000
2	0.142	0.318	0.000
3	0.624	−0.202	0.000
P.C %	29.51	5.19	0.00
LOADING P.C %	34.70		

HAVE BEEN VARIMAX ROTATED			
	FACTOR 1	FACTOR 2	FACTOR 3
1	0.625	0.314	0.000
2	0.043	0.346	0.000
3	0.656	−0.011	0.000
P.C %	27.42	7.27	0.00
LOADING P.C %	34.70		

KMO = 0.4930
Bartlett = 0.0000

Rasch-GZ also used KMO, i.e., the Kaiser-Meyer-Olkin measure, to assess the suitability of data for factor analysis. The KMO statistic generated by Rasch-GZ ranges from 0 to 1, with the following guidelines for universal interpretation:

- **KMO ≥ 0.90**: Excellent
- **0.80 ≤ KMO < 0.90:** Good
- **0.70 ≤ KMO < 0.80:** Mediocre
- **0.60 ≤ KMO < 0.70:** Average

- **0.50 ≤ KMO < 0.60:** Poor
- **KMO < 0.50:** Unacceptable

A higher KMO value indicates that a factor analysis is likely to be useful with the data set, as it suggests that the variables share common factors. On the contrary, a low KMO value tells users that the data are not suitable for factor analysis and that the variables may not have enough common variance. In the present case, KMO = 0.4930 indicates a much lower KMO value, suggesting that the data are not fit for factor analysis and that overall adjustment of test item quality is imperative. The common practice for this includes moderating poor test items, deleting invalid items or options by replacing with well-calibrated ones.

In the context of item analysis, Bartlett = 0.000 typically refers to the p-value, indicating the probability of observing the test results under the null hypothesis (H_0) that variables are independent of each other. The alternative hypothesis (H_1) is that the correlation coefficient matrix is not the identity matrix, indicating the existence of correlation between variables. Therefore, a p-value of 0.000 (often reported as < 0.001) tells the users that the result is statistically significant. This means there exists strong evidence against the null hypothesis, leading to the conclusion that there are significant differences in the variances among the groups being compared. In the present case, Bartlett = 0.0000, null hypothesis should be rejected, accepting possible correlation(s) between variables and suitable for factor analysis. Conversely, if the P-value is reported > 0.05, the null hypothesis cannot be rejected. Instead, it is considered that the independence between variables is strong and not suitable for factor analysis.

In the use of Rasch-GZ for item analysis, like other software programs, Bartlett's sphericity test is combined with KMO test. The KMO test evaluates the correlation and partial correlation between variables. The closer the KMO value is to 1, the stronger the correlation between variables, and the better the effect of factor analysis. It is generally believed that the effect is better when the KMO value is above 0.7, and factor analysis is not suitable when it is below 0.5.

Conclusion: Therefore, when the P-value of Bartlett's sphericity test is 0.0000, it indicates a significant correlation between variables and is suitable for factor analysis. However, even if the Bartlett test passes, and the KMO value is low (such as 0.4930 shown above), the quality and applicability of the data still need to be carefully considered, and data preprocessing or other dimensionality reduction methods may need to be chosen. What users need to do in the present case is to conduct the overall adjustment of test item quality. This includes moderating the poor test items (as those listed in Listening Comprehension Section, Grammar Section, and Reading Comprehension section), deleting invalid items or options by replacing with well-calibrated ones and to administer the test again.

Appendix V

Rasch Model, Logistic Normal Metric (D=1.7)

Item Difficulties (B)			
Number	Item NO.	Difficulty	Standard Error(di)
1	1	−2.024	0.224
2	2	0.453	0.260
3	3	−0.266	0.217
4	4	−0.359	0.214
5	5	−0.747	0.204
6	6	0.595	0.272
7	7	0.453	0.260
8	8	0.595	0.272
9	9	0.034	0.231
10	10	−0.071	0.226
11	11	−0.359	0.214
12	12	−0.404	0.212
13	13	0.453	0.260
14	14	−0.359	0.214
15	15	−0.404	0.212
16	16	0.386	0.254
17	17	−0.171	0.221
18	18	0.595	0.272
19	19	−0.019	0.229
20	20	0.453	0.260
21	21	0.522	0.265
22	22	−0.019	0.229
23	23	−0.359	0.214
24	24	−0.579	0.207
25	25	0.144	0.238
26	26	0.671	0.279
27	27	0.034	0.231
28	28	−0.219	0.219
29	29	−0.121	0.223
30	30	−0.121	0.223
31	31	−0.579	0.207
32	32	−1.113	0.202
33	33	−0.019	0.229
34	34	*	*

Item Difficulties (B)			
Number	Item NO.	Difficulty	Standard Error(di)
35	35	0.088	0.234
36	36	0.201	0.241
37	37	−0.071	0.226
38	38	0.088	0.234
39	39	0.261	0.245
40	40	0.034	0.231
41	41	0.144	0.238
42	42	−0.404	0.212
43	43	−0.019	0.229
44	44	0.201	0.241
45	45	0.201	0.241
46	46	0.386	0.254
47	47	0.201	0.241
48	48	0.261	0.245
49	49	−0.019	0.229
50	50	−0.121	0.223
51	51	0.144	0.238
52	52	−0.071	0.226
53	53	−0.171	0.221
54	54	−0.266	0.217
55	55	−0.019	0.229
56	56	0.034	0.231
57	57	0.144	0.238
58	58	−0.071	0.226
59	59	0.386	0.254
60	60	0.144	0.238
61	61	−0.219	0.219
62	62	0.088	0.234
63	63	−0.171	0.221
64	64	−0.121	0.223
65	65	0.261	0.245
66	66	−0.493	0.210
67	67	−0.071	0.226
68	68	0.144	0.238
69	69	−0.219	0.219
70	70	0.261	0.245
71	71	0.034	0.231
72	72	0.088	0.234

Item Difficulties (B)			
Number	**Item NO.**	**Difficulty**	**Standard Error(di)**
73	73	−0.313	0.215
74	74	−0.121	0.223
75	75	0.088	0.234
76	76	0.322	0.250
77	77	0.201	0.241
78	78	0.201	0.241
79	79	0.088	0.234
80	80	0.201	0.241
81	81	0.088	0.234
82	82	−0.121	0.223
83	83	−0.019	0.229
84	84	0.261	0.245
85	85	0.522	0.265
86	86	0.595	0.272
87	87	0.595	0.272
88	88	0.929	0.305
89	89	0.386	0.254
90	90	−0.493	0.210
91	91	−0.071	0.226
92	92	−0.266	0.217
93	93	−0.019	0.229
94	94	0.671	0.279
95	95	−0.171	0.221
96	96	−0.071	0.226
97	97	0.201	0.241
98	98	−0.121	0.223
99	99	−0.911	0.202
100	100	−0.970	0.202

All Possible Scores on the Test		
Corr Number	**Score/Ability**	**Standard Error (br)**
1	**−4.595**	**0.962**
2	**−3.901**	**0.691**
3	**−3.490**	**0.571**
4	**−3.193**	**0.499**
5	**−2.961**	**0.451**
6	**−2.768**	**0.415**
7	**−2.603**	**0.388**

All Possible Scores on the Test		
Corr Number	**Score/Ability**	**Standard Error (br)**
8	−2.457	0.365
9	−2.328	0.347
10	−2.211	0.332
11	−2.104	0.319
12	−2.005	0.308
13	−1.912	0.298
14	−1.826	0.289
15	−1.744	0.281
16	−1.667	0.274
17	−1.593	0.268
18	−1.523	0.262
19	−1.456	0.257
20	−1.391	0.253
21	−1.329	0.248
22	−1.268	0.244
23	−1.210	0.241
24	−1.153	0.237
25	−1.098	0.234
26	−1.044	0.232
27	−0.992	0.229
28	−0.941	0.226
29	−0.891	0.224
30	−0.842	0.222
31	−0.793	0.220
32	−0.746	0.218
33	−0.699	0.217
34	−0.653	0.215
35	−0.608	0.214
36	−0.563	0.213
37	−0.519	0.211
38	−0.476	0.210
39	−0.432	0.209
40	−0.389	0.209
41	−0.347	0.208
42	−0.305	0.207
43	−0.263	0.207
44	−0.221	0.206
45	−0.179	0.206
46	−0.138	0.205

All Possible Scores on the Test		
Corr Number	Score/Ability	Standard Error (br)
47	−0.097	0.205
48	−0.056	0.205
49	−0.015	0.205
50	0.027	0.205
51	0.068	0.205
52	0.109	0.205
53	0.150	0.205
54	0.191	0.205
55	0.233	0.206
56	0.274	0.206
57	0.316	0.207
58	0.358	0.207
59	0.401	0.208
60	0.443	0.209
61	0.486	0.210
62	0.530	0.211
63	0.574	0.212
64	0.618	0.213
65	0.663	0.215
66	0.709	0.216
67	0.755	0.218
68	0.802	0.219
69	0.850	0.221
70	0.898	0.223
71	0.948	0.225
72	0.998	0.228
73	1.050	0.230
74	1.103	0.233
75	1.158	0.236
76	1.214	0.239
77	1.271	0.243
78	1.331	0.247
79	1.393	0.251
80	1.456	0.256
81	1.523	0.261
82	1.592	0.266
83	1.665	0.272
84	1.741	0.279
85	1.821	0.287
86	1.907	0.296

All Possible Scores on the Test		
Corr Number	**Score/Ability**	**Standard Error (br)**
87	1.998	0.305
88	2.095	0.317
89	2.201	0.330
90	2.317	0.345
91	2.444	0.363
92	2.588	0.385
93	2.751	0.413
94	2.941	0.448
95	3.172	0.497
96	3.465	0.568
97	3.874	0.689
98	4.658	1.006

Appendix V lists the difficulty parameters (in logit) based on the 100 test items used as **Basal Test** before equating is conducted.

Interested readers may check with the difficulty parameters obtained from another 100 test items used as the **Test to Be Equated** in Appendix VI.

Only with these two sets of difficulty parameters connected with 30 linking items would the test equating be possible. Table 5.2 lists the difficulties of the two parameter sets (after equating was conducted) are comparable on the same scale (see Chapter 5 for details).

Note 1. Item Number(s) Deleted: [34], Reason for deletion: All correct answers.

Note 2. All possible scores on the test refer to all the abilities of test takers taking the test.

Note 3. In two cases, no score or ability estimate will be given: The ability having correct answers to all the test items or the ability getting no correct answers to all items, in either of which case the ability cannot be (ML) calibrated; therefore, 100 items minus one (all correct) = 99 items. Only the ability having 98 correct answers can be calibrated.

Apart from the difficulty and ability tables above, distributions of both difficulty and ability are also generated. Figure A.1 shows the item difficulty distribution wherein the horizontal axis represents the range of item difficulties, and the vertical axis represents the number of test items. The height of the bar chart reflects the actual concentration level of the test items in each difficulty range. And the dotted line shows the normal distribution for reference.

In Figure A.2, the output file with –rsh ending shows the ability distribution of all the possible test takers where the horizontal axis represents the range of ability estimates, and the vertical axis represents the number of test takers. The height of the bar chart reflects the actual level inherent in each

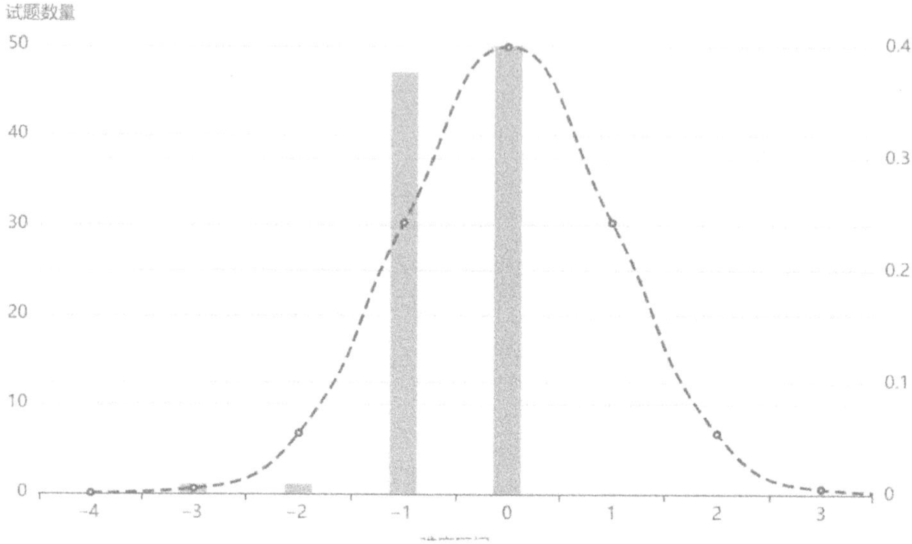

FIGURE A.1
The distribution of item difficulties of a whole test paper

ability within all and possible ability range. And the dotted curve shows the normal distribution for reference.

Figure A.3 is the scatter plot visually displaying the test item difficulty distribution. The numbers in the graph represent the item numbers in the test paper, and the size is proportional to the absolute value of difficulty.

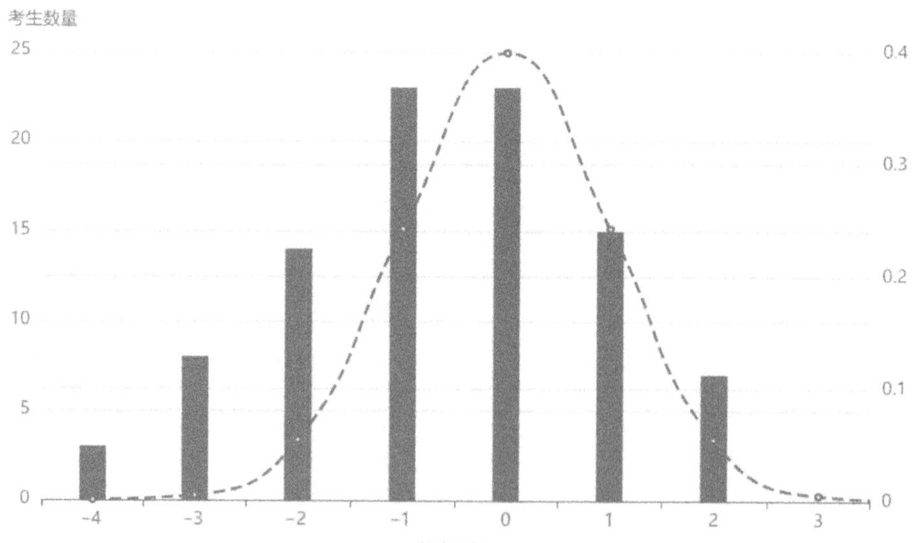

FIGURE A.2
The distribution of ability estimates based on all and possible test takers

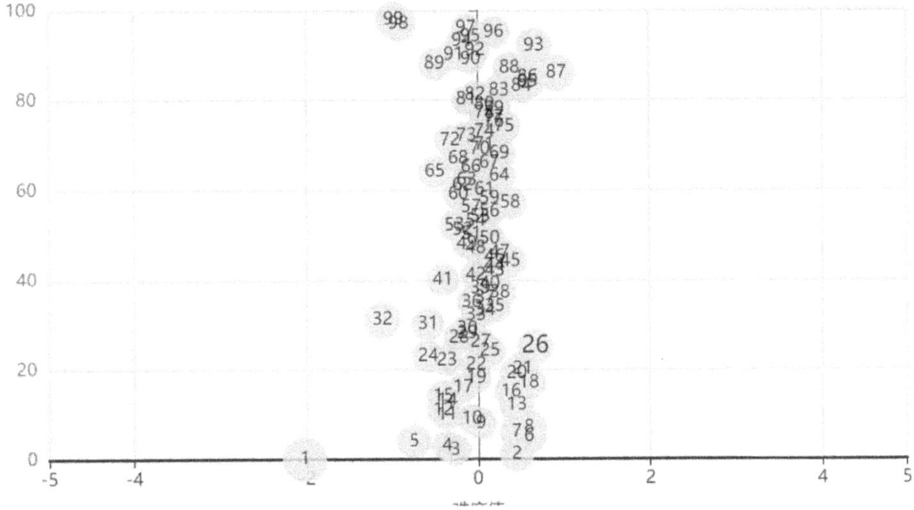

FIGURE A.3
The scatter plot of test item difficulties

Figure A.4 shows the ability distribution. The curve best displays the distribution characteristics of test taker ability. The connecting points on the curve represent ability estimates corresponding to different numbers of correct answers.

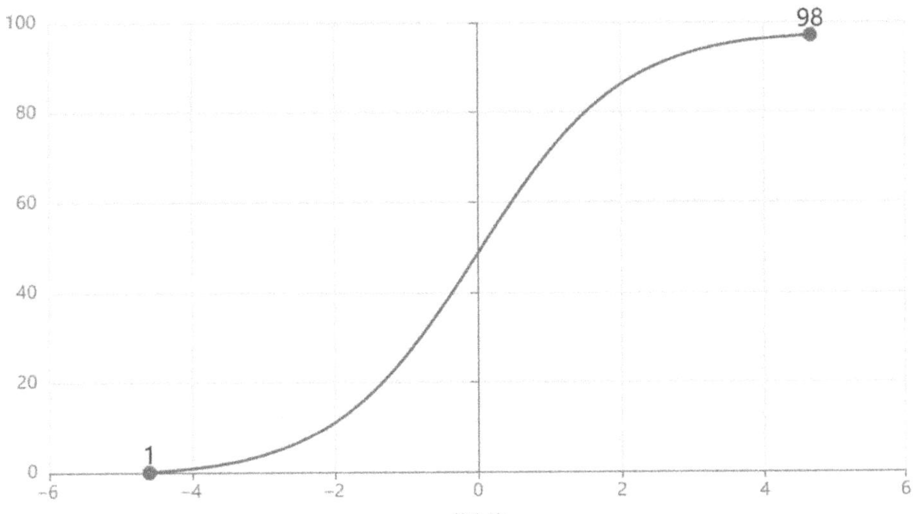

FIGURE A.4
The curve of test taker ability

Appendix VI

Rasch Model, 1 Parameter Logistic Normal METRIC (D=1.7)

TEST NAME: equated

TEST DATE: 2025-06-03

Appendix VI lists the difficulty parameters (in logit) based on another 100 test items used as the **Test to Be Equated.** Interested readers may refer back to the 100 difficulty parameters used as the **Basal Test** in Appendix V.

The difficulties of the two parameter sets here (before equating was conducted) are not comparable on the same scale. Table 5.2 lists the difficulties of these two parameter sets (after equating was conducted) are comparable on the same scale (see Chapter 5 in details).

Item Difficulties (B)			
Number	**Item NO.**	**Difficulty**	**Standard Error(di)**
1	1	−2.021	0.224
2	2	0.449	0.259
3	3	−0.268	0.217
4	4	−0.536	0.208
5	5	0.833	0.295
6	6	0.591	0.271
7	7	−0.622	0.206
8	8	0.591	0.271
9	9	0.032	0.231
10	10	0.086	0.234
11	11	−0.268	0.217
12	12	−0.493	0.209
13	13	0.449	0.259
14	14	0.591	0.271
15	15	0.319	0.249
16	16	0.319	0.249
17	17	−0.172	0.221
18	18	0.258	0.245
19	19	0.519	0.265
20	20	0.748	0.286
21	21	−1.071	0.202
22	22	−0.021	0.228
23	23	0.086	0.234
24	24	−0.579	0.207

Item Difficulties (B)			
Number	Item NO.	Difficulty	Standard Error(di)
25	25	0.141	0.237
26	26	0.667	0.278
27	27	0.032	0.231
28	28	−0.220	0.219
29	29	−0.123	0.223
30	30	−0.123	0.223
31	31	−0.579	0.207
32	32	−1.112	0.202
33	33	−0.021	0.228
34	34	−3.389	0.349
35	35	0.141	0.237
36	36	0.199	0.241
37	37	−0.073	0.226
38	38	0.086	0.234
39	39	0.258	0.245
40	40	0.032	0.231
41	41	0.141	0.237
42	42	−0.405	0.212
43	43	−0.021	0.228
44	44	0.199	0.241
45	45	0.199	0.241
46	46	0.383	0.254
47	47	0.199	0.241
48	48	−0.172	0.221
49	49	0.199	0.241
50	50	0.591	0.271
51	51	0.141	0.237
52	52	0.833	0.295
53	53	−0.172	0.221
54	54	−0.073	0.226
55	55	−0.021	0.228
56	56	0.319	0.249
57	57	0.383	0.254
58	58	−0.073	0.226
59	59	−0.449	0.211
60	60	0.199	0.241
61	61	−0.220	0.219
62	62	0.086	0.234

Item Difficulties (B)			
Number	**Item NO.**	**Difficulty**	**Standard Error(di)**
63	63	0.319	0.249
64	64	−0.073	0.226
65	65	−0.123	0.223
66	66	0.591	0.271
67	67	−0.073	0.226
68	68	0.258	0.245
69	69	−0.220	0.219
70	70	−0.405	0.212
71	71	0.032	0.231
72	72	0.086	0.234
73	73	−0.314	0.215
74	74	−0.123	0.223
75	75	0.086	0.234
76	76	0.319	0.249
77	77	0.199	0.241
78	78	0.199	0.241
79	79	0.086	0.234
80	80	0.199	0.241
81	81	0.086	0.234
82	82	−0.123	0.223
83	83	−0.021	0.228
84	84	0.258	0.245
85	85	0.519	0.265
86	86	0.591	0.271
87	87	0.591	0.271
88	88	0.924	0.305
89	89	0.383	0.254
90	90	−0.493	0.209
91	91	−0.073	0.226
92	92	−0.268	0.217
93	93	−0.021	0.228
94	94	0.667	0.278
95	95	−0.172	0.221
96	96	−0.073	0.226
97	97	0.199	0.241
98	98	−0.123	0.223
99	99	−0.910	0.202
100	100	−0.969	0.202

All Possible Scores on the Test		
Corr Number	Score/Ability	Standard Error(br)
1	−4.781	0.978
2	−4.060	0.708
3	−3.627	0.587
4	−3.314	0.514
5	−3.068	0.464
6	−2.864	0.427
7	−2.690	0.399
8	−2.537	0.375
9	−2.400	0.356
10	−2.277	0.340
11	−2.164	0.326
12	−2.061	0.315
13	−1.964	0.304
14	−1.874	0.295
15	−1.789	0.287
16	−1.709	0.280
17	−1.632	0.273
18	−1.559	0.267
19	−1.490	0.262
20	−1.423	0.257
21	−1.359	0.252
22	−1.296	0.248
23	−1.236	0.244
24	−1.178	0.241
25	−1.122	0.237
26	−1.066	0.234
27	−1.013	0.232
28	−0.960	0.229
29	−0.909	0.227
30	−0.859	0.224
31	−0.810	0.222
32	−0.761	0.221
33	−0.714	0.219
34	−0.667	0.217
35	−0.621	0.216
36	−0.575	0.214
37	−0.531	0.213
38	−0.486	0.212

All Possible Scores on the Test		
Corr Number	Score/Ability	Standard Error(br)
39	−0.442	0.211
40	−0.399	0.210
41	−0.356	0.209
42	−0.313	0.208
43	−0.271	0.208
44	−0.228	0.207
45	−0.186	0.207
46	−0.145	0.206
47	−0.103	0.206
48	−0.062	0.206
49	−0.020	0.205
50	0.021	0.205
51	0.062	0.205
52	0.104	0.205
53	0.145	0.205
54	0.186	0.206
55	0.228	0.206
56	0.270	0.206
57	0.311	0.207
58	0.354	0.207
59	0.396	0.208
60	0.438	0.209
61	0.481	0.209
62	0.525	0.210
63	0.568	0.211
64	0.612	0.212
65	0.657	0.214
66	0.702	0.215
67	0.748	0.216
68	0.795	0.218
69	0.842	0.220
70	0.890	0.222
71	0.939	0.224
72	0.989	0.226
73	1.039	0.228
74	1.092	0.231
75	1.145	0.233
76	1.200	0.236
77	1.256	0.240

All Possible Scores on the Test		
Corr Number	**Score/Ability**	**Standard Error(br)**
78	1.314	0.243
79	1.373	0.247
80	1.435	0.251
81	1.499	0.256
82	1.566	0.261
83	1.636	0.267
84	1.708	0.273
85	1.785	0.280
86	1.866	0.287
87	1.951	0.296
88	2.042	0.306
89	2.140	0.317
90	2.246	0.330
91	2.362	0.345
92	2.490	0.363
93	2.634	0.385
94	2.797	0.413
95	2.988	0.448
96	3.218	0.497
97	3.512	0.568
98	3.921	0.689
99	4.706	1.006

Difficulty Distribution of Test Items

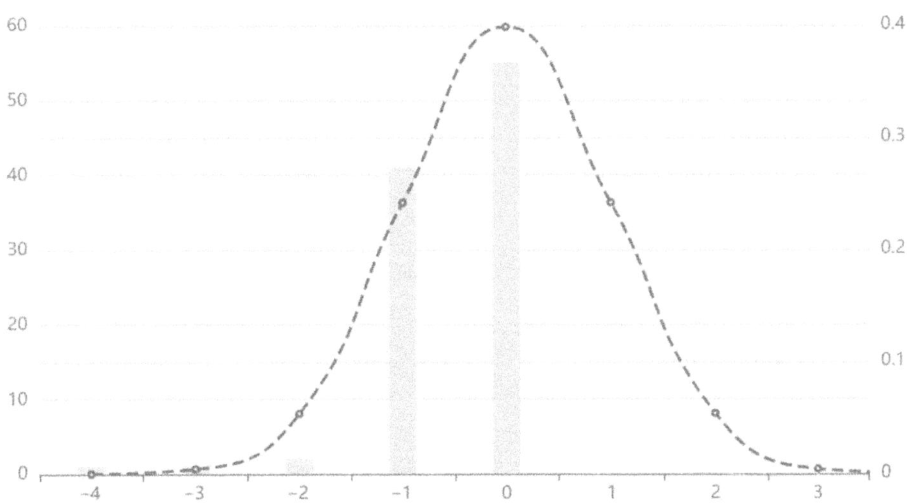

Ability Distribution of Test Takers

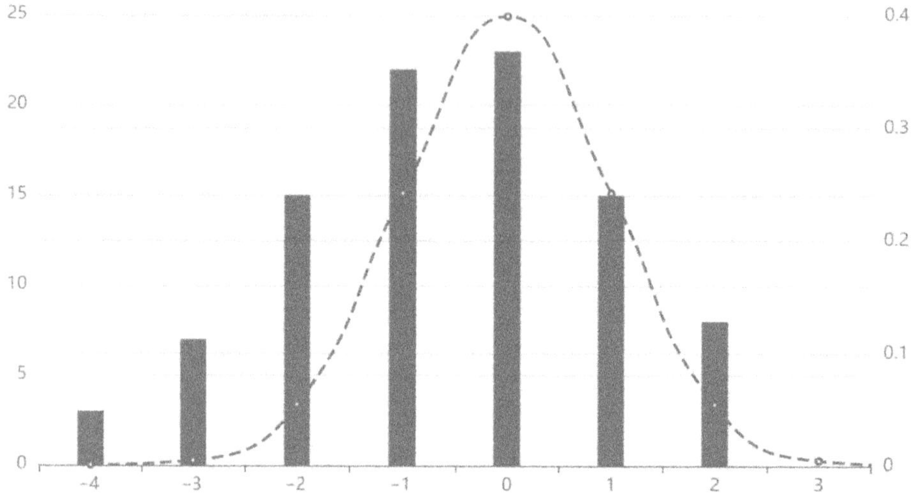

The Chart of Test Item Difficulty (The Wright Map)

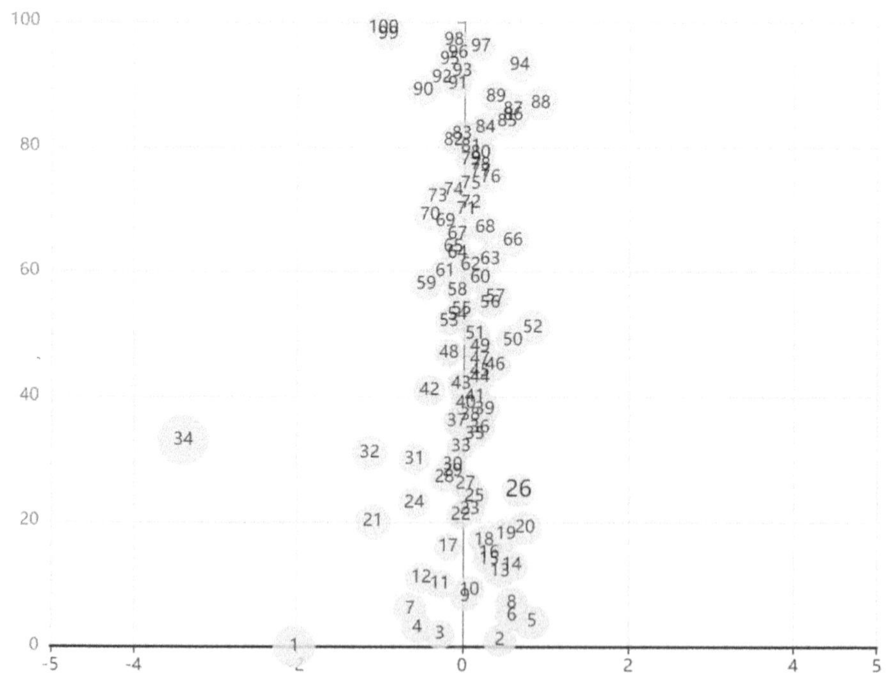

This is the distribution of the test item difficulty.

(1) The numbers in the chart represent the specific test items;
(2) The horizontal axis shows the difficulty range;
(3) The difficulty of all correct or all incorrect items is not within the estimated range.

The Curve of Test Taker Ability

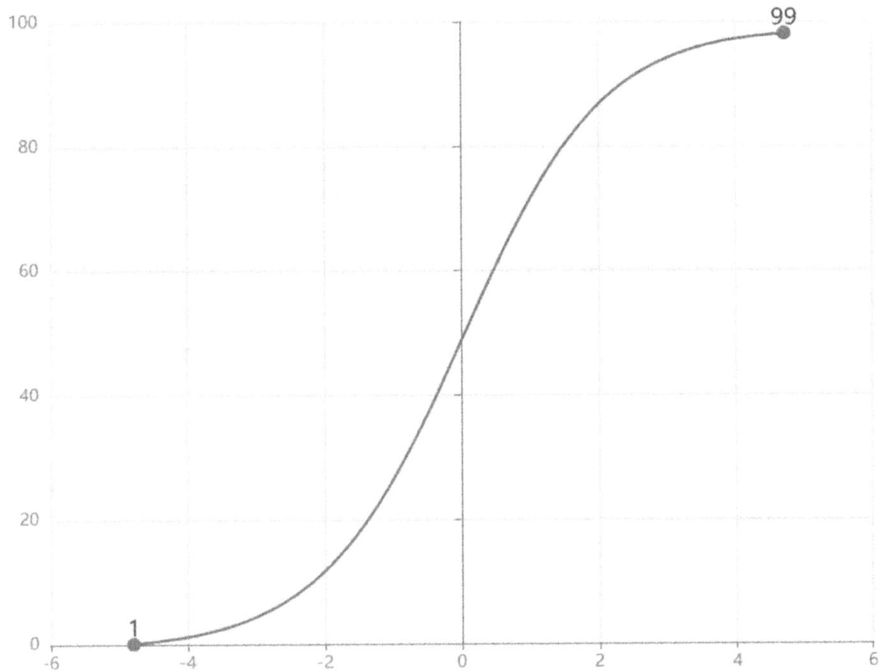

This is the ability curve of the test takers in this test.

(1) It indicates the ability estimate of answering 1 test item correctly to (total number - 1) test items correctly;
(2) The numbers in the chart represent the number of correct answers;
(3) The ability estimates of test takers who answered all the test items correctly or incorrectly are not included in the estimation range.

— End Rasch-GZ

5

RASCH-GZ: Basic Operating Steps for Test Equating

In Chapter 3, we introduced the conditions that should be met for test equating, specifically mentioning the need to set a certain test paper as a basal test and prepare other parallel test papers that need to be equated. We also emphasize that test items in both basal test paper and the parallel test papers to be equated need to undergo strict calibration via item analysis so as to obtain the file with –rsh extension containing both item difficulty parameters and corresponding ability estimates based on the corresponding number of correct answers. With the "-rsh files", we can feasibly conduct test equating.

This chapter will present, via illustration of either screenshots, figures, or tables, step by step, the basic operating steps of the Rasch GZ system for test equating. Meanwhile, readers can also download the short videos guider to the Rasch GZ system to obtain the most direct and intuitive demonstration of the complete procedure for test equating.

5.1 The Rasch GZ System: Steps for Test Equating

At this very outset, users must know the data files required for test equating.

To conduct test equating using the Rasch-GZ, the following data files must be available:

(1) At least two data files with "–rsh" extension (2.3) are required. One file is set as the basal test, while the other as the one to be equated. In our example, the output data file of Appendix III is used as the basal test and that of Appendix IV is used as the test to be equated;

(2) The format of these two test papers should be the same, but the number of test items may or may not be exactly the same;

(3) A set of items used as the linking items must be specified and put in both tests. In this example, 30 items were specified as linking items and put in both data files.

To conduct test equating via the Rasch-GZ, users do not need to create any command files or control files as required by most of the software packages

DOI: 10.1201/9781003665595-5

FIGURE 5.1
Screenshot of the Rasch GZ system test equating window

(see Section 3.8). Just follow the steps by clicking the mouse to enter the test equating interface to begin.

Step 1: We enter the test equating interface. Figure 5.1 shows a screenshot of the Rasch GZ test equating window.

Step 2: Follow the prompts in the above figure and click the "Open File" button in the box of Test 1 (Basal Test). The system will pop up a list of data files, and select the data file with the extension "- rsh" as the basal file from it. Please note that the data files at this point must all end with "-. rsh", indicating the files contain difficulty parameter/ability estimates of the test items. The test paper name, number of test items, and number of test takers will be displayed in the basal test information box as shown in Figure 5.2. If there are no data files with the extension "- rsh" available, users should go back to the previous phase to conduct the Rasch analysis calibration to generate the data files with the extension "- rsh" as shown in Figure 4.10. Click the choice indicating "–rsh" to obtain the file with extension of "–rsh".

Step 3: Select the data information for the test to be equated. Click the "Open File" button in the second box of the test paper, and the system will pop up a list of data files. Select the data file with - rsh as the extension, that is, the data file with the difficulty parameters/ability estimates. The name of the test paper, the number of test item, and the number of test takers will be displayed in the information box as shown in Figure 5.3.

Step 4: Conduct Chi Square Test on Linking Items

At this point, the users are supposed to set the exact number of linking items. In the "Set Number of Linking Items" input box, enter the number of linking items. Click the minus sign (-) on the left to decrease the quantity in

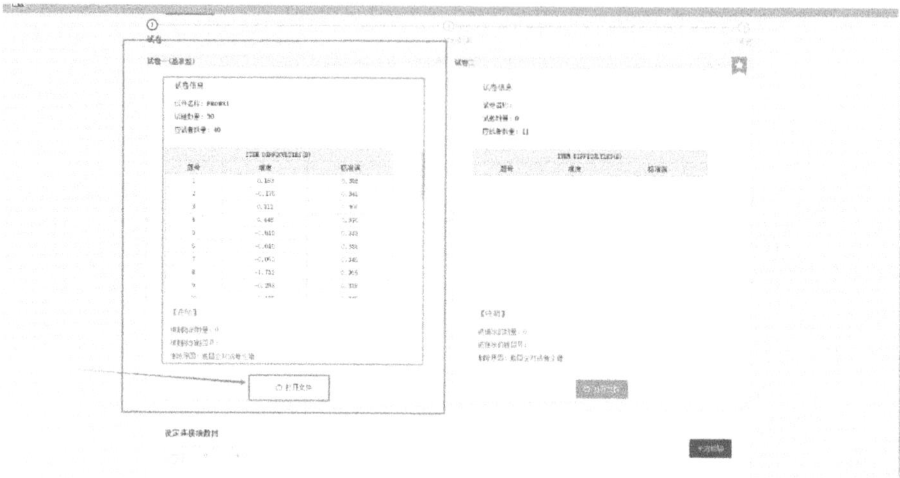

FIGURE 5.2
Screenshot of test equating interface of the Rasch GZ system for Basal test

sequence, and click the plus sign (+) on the right to increase the quantity in sequence (as shown in Figure 5.4). This is a new feature of the RASCH-GZ, which allows users to independently determine the number of linking items prior to running the program, unlike other software packages that require

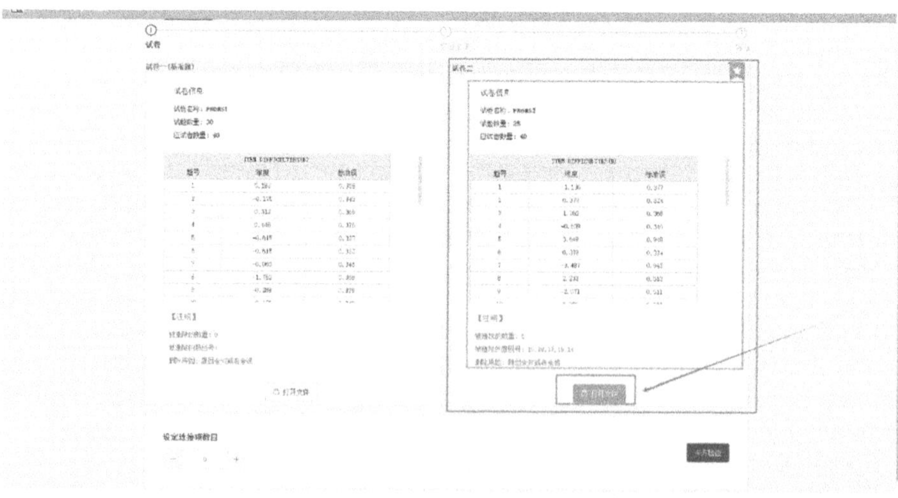

FIGURE 5.3
Screenshot of test equating interface of the Rasch GZ system for test to be equated

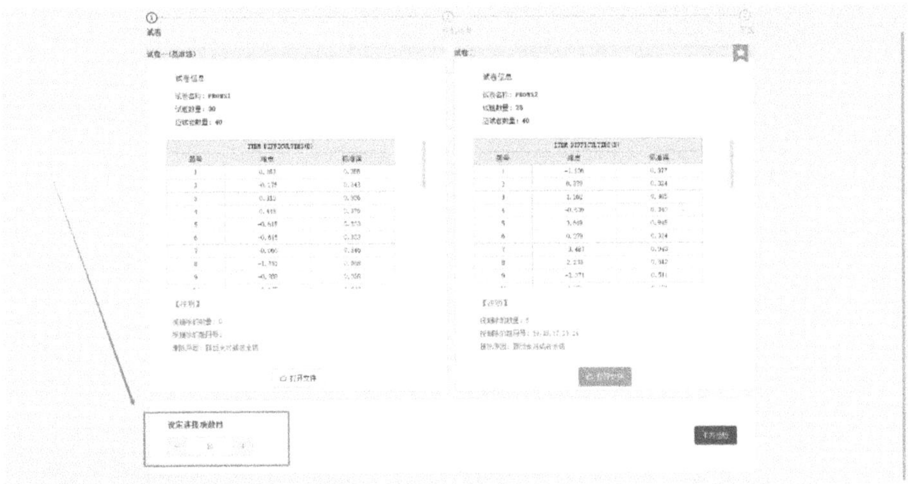

FIGURE 5.4
Screenshot of Rasch GZ system performing Chi square test on linking items

setting in the command file before running the program or/and exiting the program to enter the command file for specific modification.

Another point to note is that when using linking items for test equating, theoretically these linking items can be placed in various parts of a test paper. However, for better data editing purpose, the test items used as linking items are generally placed at the beginning of the data file for ease of program reading.

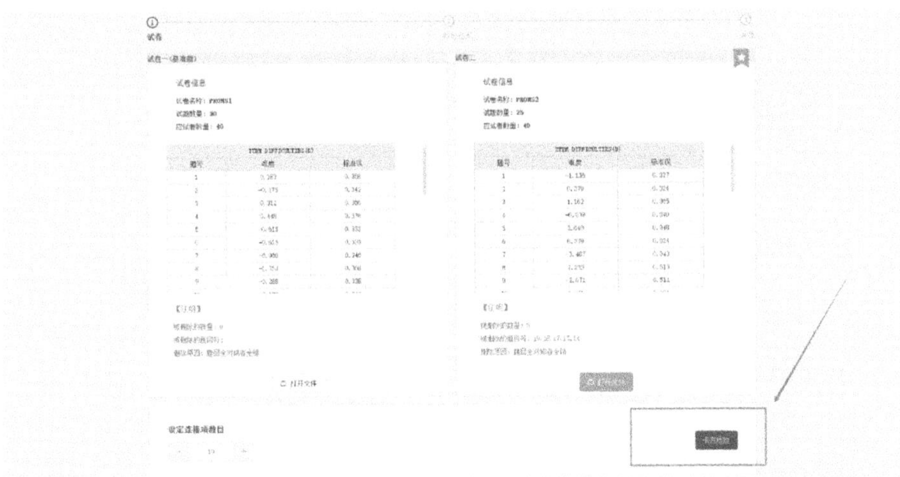

FIGURE 5.5
Screenshot of the Rasch GZ system linking item settings interface

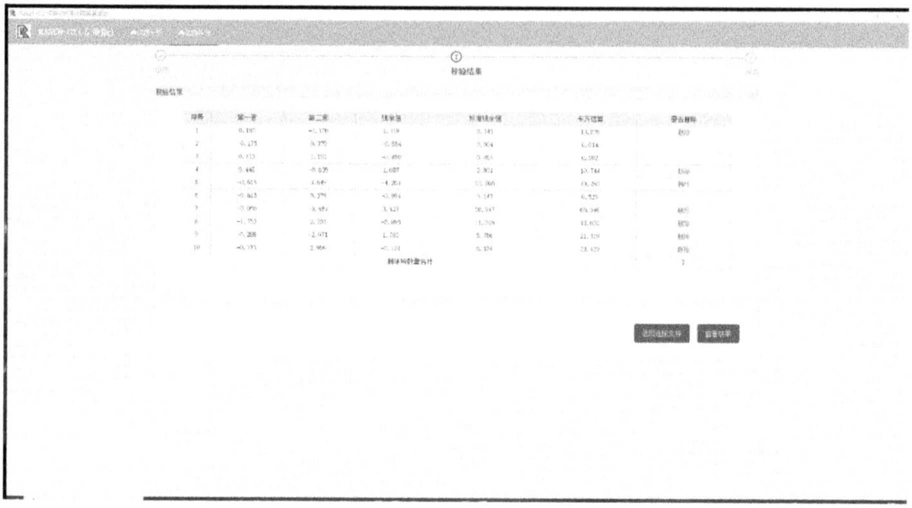

FIGURE 5.6
The interface of Chi square test result based on linking items

Then click the "Chi Square Test" Button (as shown in Figure 5.5) to view the test results.

Step 5: Display the Chi Square Test Results of the Linking Items

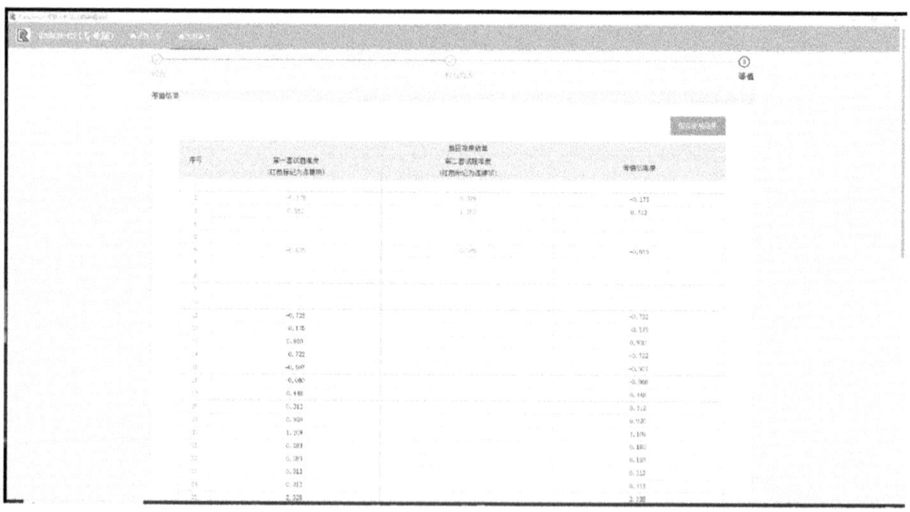

FIGURE 5.7
Screenshot of Test Equating Results generated by The Rasch GZ System

The RASCH-GZ system automatically conducts Chi Square Test on the linking items for both the basal test paper and the test paper to be equated. After the Chi SquareTest, the linking items that do not meet the criteria will be marked as "deleted" and highlight in red, and the all the other linking items that meet the criteria will be automatically used in the test equating process. Figure 5.6 shows the Chi Square Test result interface of the Rasch GZ system.

Step 6: Display the Test Equating Results

Click the "Equating Results" button on the toolbar, and the Rasch-GZ system will automatically conduct the test equating on the basal test and the test to be equated using the specified linking items (i.e., those that pass the Chi square test). The output file of test equating ends with -.eqt. Figure 5.7 shows the screenshot of test equating results.

5.2 Interpretation of Test Equating Results

At this very outset, we should understand that the test difficulty parameters listed in Appendix III and Appendix IV are not yet comparable on the same scale. By now, they are two independent output files with ending –rsh, each with the same 30 linking items. Table 5.1 shows the linking item difficulties obtained, respectively, from the two output files.

Table 5.1 gives the difficulty parameters of the 30 linking items obtained from two different tests. What we are interested in here is the difficulty means of the 30 linking items in the two sets of tests. As observed in the above table, the two separate means based on the same linking items turned out to be different: −0.04 (logits) in the basal test and 0.006 (logits) in the test to equate. Our question is: Why are the difficulties of the same 30 items different? Our interpretation is that the overall difficulty level of the two test items to which the 30 linking items are connected is different. If −0.04 – (0.006), the difference thus obtained from the Mean1 minus the Mean 2 is −0.047 logit, indicating the test items in basal test are a little bit difficult than those in the test to be equated. That is why the means of the 30 linking items in the basal test turn out to be easier than those in the test to be equated. In other words, test items in the basal test are 0.047 difficult in logit than those in the test to be equated. To quote Ben Wright, "In such an example, the linking items are the hard items in EASY test but the easy items in the HARD test" (Wright and Stone, 1979; Wright, 1992; Gui et al., 1993).

As a general practice, we equate all the test items based on the basal test to obtain the difficulty parameters as shown in Table 5.2. At this point, the

TABLE 5.1

Linking Item Difficulties Obtained from the Two Output Files

Linking Items in Basal Test			Linking Items in Test to be Equated		
Item No.	**Difficulty**	**Standard Error (di)**	**Item No.**	**Difficulty**	**Standard Error (di)**
1	−2.024	0.224	1	−2.021	0.224
2	0.453	0.260	2	0.449	0.259
3	−0.266	0.217	3	−0.268	0.217
4	−0.359	0.214	4	−0.536	0.208
5	−0.747	0.204	5	0.833	0.295
6	0.595	0.272	6	0.591	0.271
7	0.453	0.260	7	−0.622	0.206
8	0.595	0.272	8	0.591	0.271
9	0.034	0.231	9	0.032	0.231
10	−0.071	0.226	10	0.086	0.234
11	−0.359	0.214	11	−0.268	0.217
12	−0.404	0.212	12	−0.493	0.209
13	0.453	0.260	13	0.449	0.259
14	−0.359	0.214	14	0.591	0.271
15	−0.404	0.212	15	0.319	0.249
16	0.386	0.254	16	0.319	0.249
17	−0.171	0.221	17	−0.172	0.221
18	0.595	0.272	18	0.258	0.245
19	−0.019	0.229	19	0.519	0.265
20	0.453	0.260	20	0.748	0.286
21	0.522	0.265	21	−1.071	0.202
22	−0.019	0.229	22	−0.021	0.228
23	−0.359	0.214	23	0.086	0.234
24	−0.579	0.207	24	−0.579	0.207
25	0.144	0.238	25	0.141	0.237
26	0.671	0.279	26	0.667	0.278
27	0.034	0.231	27	0.032	0.231
28	−0.219	0.219	28	−0.220	0.219
29	−0.121	0.223	29	−0.123	0.223
30	−0.121	0.223	30	−0.123	0.223
Mean 1	−0.04		Mean 2	0.006	

TABLE 5.2

The Equated Test Item Difficulties Which Can Be Comparable on the Same Scale

No.	Item difficulties rescaled and equated comparable on the same scale
1	−2.024
2	0.453
3	−0.266
4	−0.359
5	*
6	0.595
8	0.595
7	*
9	0.034
10	−0.071
11	−0.359
12	−0.404
13	0.453
16	0.386
17	−0.171
14	*
18	0.595
15	*
19	−0.019
20	0.453
22	−0.019
23	−0.359
24	−0.579
25	0.144
21	*
26	0.671
27	0.034
28	−0.219
29	−0.121
30	−0.121
31	−0.579
32	−1.113
33	−0.019
35	0.088

(*Continued*)

TABLE 5.2 (***Continued***)

The Equated Test Item Difficulties Which Can Be Comparable on the Same Scale

No.	Item difficulties rescaled and equated comparable on the same scale
36	0.201
37	−0.071
38	0.088
39	0.261
34	*
40	0.034
41	0.144
42	−0.404
43	−0.019
44	0.201
45	0.201
46	0.386
47	0.201
48	0.261
49	−0.019
50	−0.121
51	0.144
52	−0.071
53	−0.171
54	−0.266
55	−0.019
56	0.034
57	0.144
58	−0.071
59	0.386
60	0.144
61	−0.219
62	0.088
63	−0.171
64	−0.121
65	0.261
66	−0.493
67	−0.071
68	0.144

TABLE 5.2 ***(Continued)***

The Equated Test Item Difficulties Which Can Be Comparable on the Same Scale

No.	Item difficulties rescaled and equated comparable on the same scale
69	−0.219
70	0.261
71	0.034
72	0.088
73	−0.313
74	−0.121
75	0.088
76	0.322
77	0.201
78	0.201
79	0.088
80	0.201
81	0.088
82	−0.121
83	−0.019
84	0.261
85	0.522
86	0.595
87	0.595
88	0.929
89	0.386
90	−0.493
91	−0.071
92	−0.266
93	−0.019
94	0.671
95	−0.171
96	−0.071
97	0.201
98	−0.121
99	−0.911
100	−0.97
101	−0.612
102	−1.145

(Continued)

TABLE 5.2 (*Continued*)

The Equated Test Item Difficulties Which Can Be Comparable on the Same Scale

No.	Item difficulties rescaled and equated comparable on the same scale
103	−0.054
104	−3.422
105	0.108
106	0.166
107	−0.106
108	0.053
109	0.225
110	−0.001
111	0.108
112	−0.438
113	−0.054
114	0.166
115	0.166
116	0.35
117	0.166
118	−0.205
119	0.166
120	0.558
121	0.108
122	0.8
123	−0.205
124	−0.106
125	−0.054
126	0.286
127	0.35
128	−0.106
129	−0.482
130	0.166
131	−0.253
132	0.053
133	0.286
134	−0.106
135	−0.156
136	0.558

TABLE 5.2 (*Continued*)

The Equated Test Item Difficulties Which Can Be Comparable on the Same Scale

No.	Item difficulties rescaled and equated comparable on the same scale
137	−0.106
138	0.225
139	−0.253
140	−0.438
141	−0.001
142	0.053
143	−0.347
144	−0.156
145	0.053
146	0.286
147	0.166
148	0.166
149	0.053
150	0.166
151	0.053
152	−0.156
153	−0.054
154	0.225
155	0.486
156	0.558
157	0.558
158	0.891
159	0.35
160	−0.526
161	−0.106
162	−0.301
163	−0.054
164	0.634
165	−0.205
166	−0.106
167	0.166
168	−0.156
169	−0.943
170	−1.002

difficulty of all the test items is comparable on the same scale. Table 5.2 gives the equated test item difficulties of both tests.

In Table 5.2, the asterisked items were deleted due to being unqualified fitness during the equating process: [5, 7, 14, 15, 21, 34];

The items from No.1 to No.100 are of the basal test, and beginning from No.101 to No.170 are test items of the test being equated. Both tests have the same 30 linking items, so the total is

Linking Items 30 + 70 items (Basal Test)

Linking Items 30 + 70 items (Test being equated)

5.3 Summary

This chapter presents a complete process of test equating via the Rasch-GZ system. The examples given here illustrate good solutions for objective measurement problems encountered in language testing, giving readers a clear understanding of how test equating is conducted. The development and use of Rasch-GZ system are feasible for language testing researchers, professionals, and practitioners. The reason is obvious. The raw (dichotomous) data of language testing is difficult to control, and the concepts are also vague. In this sense, the RASCH-GZ system effectively calibrates binary data to generate quantitative parameters, greatly promoting researchers' understanding and application of the Rasch model.

In particular, the author would like to summarize four important points regarding the test equating practice. At the first place, in language testing today, equating remains the prerequisite condition for computerized adaptive testing (CAT), item banking, and the online testing under the Internet-based environment as well. With equating, the changes of item difficulties can be monitored and equated, and the corresponding ability estimates across different occasions are thus rescaled.

Second, the significance inherent in test equating is to maintain test security and fairness. It is known to all in the practice of test item design and writing, even if two parallel test papers were constructed as similar as possible in both content and forms, they still differ somewhat in difficulty and discrimination. According to Thorndike (1971) and Brian (2003), the comparability of tests scores (or ability estimates) across different tests measuring the same ability is an issue of considerable importance to test developers, measurement professionals, and test takers alike; therefore, the comparability of tests scores (or ability estimates) across different tests measuring the same ability is an important issue in language testing. In this sense, test equating plays a crucial role in examinations on large scale with high stakes such as MET and CET in China, TOEFL in the States, etc. Only the specific methods may vary and the software used has its own preferences, yet test fairness can be put well under control. Another significance contributed by the Rasch-GZ to language testing is that the results obtained justified the proposal first put forward by Wright and Stone (1979) that "the common items are the hard items in **EASY** test but the easy items in the **HARD** test". Finally, the student version of RASCH-GZ and the user guide offer the easy way for researchers to learn how to use Rasch model measurement.[1] Though processing a small data matrix, the student version offers good illustration plus short video examples to allow new users to quickly familiarize themselves with its operation steps, while facilitating classroom teaching and demonstration for graduate supervisors. In case big data collected from a large-scale examination with high stakes needs calibration, it is recommended that users register online to use the professional version of the Rasch GZ system.

To conclude, the author wants to emphasize that in today's language testing practice, what we are doing may not be necessarily more precise but must be more correct! So long as the method is correct and testing data is objective in nature, using Rasch model is the right choice because the Model can provide good solutions to many problems of objective measurement in social sciences.

Note

1. The student version of Rasch-GZ can be downloaded, free of charge, from https://www.rasch-gz.com.

Bibliography

Bond, T.G., & Fox, C.M. (2015). Applying the Rasch Model: Fundamental Measurement in the Human Science (3rd Ed). New York and London: Taylor & Francis Group.

Gui, S.C, Li, W., & Zhang, Q. (1993). The Application of IRT to MET Equating. In NEEA (Ed.), The 4th Annual Forum on Educational Testing in Beijing (pp. 391–393). Beijing, China: China Peace Press.

Hambleton, R.K., & Swaminathan, H. (1985). Item Response Theory: Principle and Application. Dordrecht, The Netherlands: Kluwer Academic Publisher. ISBN: 0-89838-063-1

Kolen, M.J., & Brennan, R.L. (1995). Test Equating: Methods and Practices. New York: Springer Verlag.

Kolen, M.J., & Brennan, R.L. (2004). Test Equating, Scaling, and Linking: Methods and Practices (2nd Ed). New York: Springer Verlag.

Liu, Y. (2020). Test Equating, Scaling, and Linking: Methods and Practices (Kolen, M.J., & Brennan, R.L., Trans.) (3rd Ed). (Original work published 2004) New York: Springer Verlag. https://doi.org/10.1007/978-1-4939-0317-7
Lynch, B.K. (2003). Language Assessment and Programme Evaluation. In Alan Davies & Keith Michel (Eds.), Edinburgh Textbooks in Applied Linguistics. Scotland, UK: Edinburgh University Press.
Thorndike, R.L. (1971). Educational Measurement (2nd Ed). USA: American Council on Education (ACE). ISBN: 978-0-8268-6205-5
Wright, B.D. (1992). IRT in the 1990s: Which models work best? Rasch Measurement Transactions, 6(1), 196–200.
Wright, B.D., Nead, R.J., & Bell, S.R. (1980). BICAL: Calibrating Items with the Rasch Model. Research Memorandum 23C. Department of Education, Chicago University.
Wright, B.D., & Stone, M.H. (1979). Best Test Design: Rasch Measurement. Chicago, IL, USA: MESA Press. ISBN: 0-941938-00-5
Zhang, Q. (2012). Towards International Practice of Language Testing in China. Keynote speech given at the PROMS2012, Jiaxing University, China, August 6–9, 2012.
Zhang, Q. (2019). Rasch Model: Research and Practice in China. In Myint (Ed.), International Trends in Educational Assessment: Emerging Issues and Practices. Leiden, The Netherlands: Koninklijke Brill NV. Retrieved from http://catalog.loc.gov
Zhang, Q. (2000). BILOG and PARSCALE: Different but Alike. In Jose Lai & Pauline Po-yee Tam (Eds.), Crosslinks in English Language Teaching. Vol. 1, 2000. English Language Teaching Unit. Hong Kong, SAR, China: The Chinese University of Hong Kong.
Zhang, Q. (2021a). Rasch-GZ, The First Chinese version of Rasch-based Item Analysis and Test Equating. In Department of Psychology, Nanjing Normal University (Chairs), Online Symposium. PROMS2021, Nanjing, China. Retrieved from http://www.proms.promsociety.org
Zhang, Q. (2021b). Rasch Model and Test Equating in China: Rasch Model in Chinese Version Has Come. In Xinling Zhang (Chairs), The 6th International Conference on Language Testing and Assessment. Shanghai, China.
Zhang, Q. (2022a). From GITEST to RASCH-GZ: Inheritance and Development of Rasch-based Research in China. http://dx.doi.org/10.20431/2349-0381.09S1001
Zhang, Q. (2022b). Rasch Model and Test Equating in China: The Most Updated Development in China. In Mariam Haji Monek (Chairs), International Education e-Colloquium, the 22nd-24th Feb. 2022. KL, Malaysia.
Zhang, Q. (2023). RASCH-GZ (Version 2.0) [Computer software]. http://www.rasch-gz.com
Zhang, Q. (2024). Rasch-GZ: The Most Updated Research Development in China. In Q. Zhang (Ed.), Proceedings of Pacific-Rim Objective Measurement Symposium (PROMS 2023) AP/Springer Nature. https://doi.org/10.2991/978-94-6463-494-5-22
张权. (2004). 《语言测试中的项目分析与等值技术: 研究与应用》. 北京: 高等教育出版社

6

Development of Objective Measurement and PROMS

6.1 Introduction to Pacific Rim Objective Measurement Symposium (PROMS)

Pacific Rim Objective Measurement Symposium (PROMS) (the symposium) is an annual non-governmental professional meeting established to promote objective measurement and contribute to the research and development of the Rasch measurement in the Pacific Rim. It is an annual symposium aiming to provide a forum for the sharing of new knowledge with the international community. The society hosts the PROMS annually and organized by local conference planning committees.

What characterizes PROMS conference can be summarized in three aspects. First, PROMS is an international and highly academic conference to be held once a year in a Pacific-Rim region or a country. The conference participants are all MA or Ph.D. candidates, Ph.D. supervisors, professionals, or practitioners engaged in psychometrics, the Rasch-based research, objective measurement, or assessment. Next, their research fields may not be the same, but they are all using the Rasch model to conduct their research. The computer software systems they used include but are not limited to WINSTEPS, FACET, R-, myRasch, GITEST, RASCH-GZ, etc. Third, pre-conference workshops of Rasch measurement are usually attached to PROMS conferences. Usually, the conference organizers arrange parallel 1–2-day pre-conference workshops of different topics to promote the knowledge and the use of the Rasch-based software. This has greatly popularized the research and application of the Rasch model for participants. In this way, PROMS conferences are believed to have provided an excellent academic platform for participants from the Pacific-rim and beyond. Therefore, today anyone seriously interested in research and development in the field of psychometrics or measurement will find such an international symposium and related workshops to be an excellent source of information about the application of the Rasch Model.

6.1.1 The First PROMS Conference

The first PROMS was held in Malaysia in 2005, as Bond recalled 20 years later, in the absence of a local organizer for the regular International Objective

DOI: 10.1201/9781003665595-6

Measurement Workshop meeting in 2004, Trevor Bond took on the organization of that IOMW XII, held at the James Cook University campus in Cairns, Australia in June 2004. The strong support of many first-time delegates from Southeast Asia changed the focus from the east of the Pacific Ocean to its western edge. Fortunately, this coincided with the intended development of research capacity at the International Islamic University of Malaysia. Prof Bond was invited to conduct a week-long Rasch research workshop at IIUM, and PROMS was born with its first meeting held in Kuala Lumpur, June 21–23, 2005.

PROMS2005 was hosted by International Islamic University of Malaysia, with Prof Noorlide Abu Kasim and Prof. Nor Akbariah Makhzir as conveners. Prof Mike Linacre conducted pre-conference workshops on WINSTEPS and Facets, and Prof. Trevor Bond gave the keynote address, "Achieving the impossible: Solving real measurement problems with the Rasch model". Later that year, Prof Bond was appointed as a Professor and Head of Department in Hong Kong Institute of Education (HKiEd), and he turned his efforts to developing and extending Rasch measurement research capacity in Eastern and South East Asia. Bond chaired the PROMS committee for more than a decade, establishing the meeting as an annual workshop and symposium in the area, with the next PROMS held in Hong Kong in 2006. The birth of PROMS laid a solid foundation for the research and application of the Rasch measurement in the Pacific Rim region and around the world.

6.1.2 The First PROMS Conference in China Mainland

Prior to going further, one very noteworthy thing is that: it is Professor Magdalena Mo Ching Mok[1] from the Educational University of Hong Kong, Hong Kong SAR, China who introduced PROMS into China in 2011. And the next year PROMS 2012 was successfully held in Jiaxing University, Zhejiang province, China from the 6th to 9th of August with pre-conference workshops scheduled on the 4th–5th of August and post-conference self-arranged events on the 10th of August, 2012.

This is the first time that PROMS conference was held in China Mainland. The convening of PROMS2012 in China can be regarded as a milestone in Rasch research in China. There are basically two reasons. The first reason is that Chinese scholars had understanding of Rasch measurement as early as in the 1980s and successfully used the Rasch model to complete the ten-year MET Equating Project (1990–1999) assigned by the Chinese Ministry of Education. MET Equating Project was recognized by foreign peers. However, due to various reasons, research in this area was basically interrupted and did not continue. This kind of suspension lasted for more than a decade. The convening of PROMS at Jiaxing University has actually rekindled the enthusiasm and pursuit of Chinese scholars for the study of Rasch models, meanwhile also strengthening the correctness of Rasch model research and people's confidence. The other reason is of practical significance. That's to say, in the environment at that time, few Chinese scholars had any information about PROMS.

Technically, among many measurement models, the Rasch model proposed by Georg Rasch is the most powerful yet feasible to use. Whether the data matrix is of dichotomous or polychotomous, they can be calibrated or scored via Rasch-based software. However, for quite a long time in China, such widely applied, powerful, and feasible methods are mostly confined within small groups of Chinese researchers of language testing who are English majors. Most Chinese scholars, particularly young Chinese non-English majored researchers, may not fully understand the ideas, concepts, applications, and efficiency inherent in the Rasch measurement methods.

Theoretically, for example, in language testing, the classical testing theory (CTT) plays the dominant role in schools and examination departments across the China mainland. This is mainly reflected in such a common practice that the difficulty of test items thus obtained is actually calibrated on the proficiency level of the test takers. In turn, the level of the test takers restricts the difficulty of test items. This has been the real situation in China's academic circles. The unique feature inherent in the Rasch model that item difficulties and subject abilities can be estimated independently of the sample is not universally known. Such an advantage needs dissemination and promotion among Chinese researchers. It is motivated by this that the present author has been making efforts in this line in China. So to speak, it is from Jiaxing that the Chinese research and development of Rasch measurement has broken through Asia into Pacific-rim and beyond.

6.1.3 The PROMS Online Conferences during the COVID-19 Pandemic Period

During the extremely severe three-year global pandemic, two PROMS conferences did not interrupt and were held in both virtual and physical setting due to the ongoing global situation at the time. One is PROMS2021 Nanjing, China, December 2021, and the other is PROMS2022 Hanoi, Vietnam, December 2–3, 2022. Both were successfully implemented on ZOOM online conference platform during the difficult time of the global fighting against the COVID-19 epidemic. The host of the former is the School of Psychology, Nanjing Normal University, China and the Jiangsu Psychological Society (Development and Educational Psychology Special Committee). The latter is the Vietnam National Institute of Educational Sciences, Hanoi, Vietnam, of which PROMS2021 held in Nanjing, China is the largest online international conference in PROMS history.

PROMS 2021 was attended by over 100 experts, scholars, and researchers in the fields of psychology and education from 11 countries and regions, including China, the United States, Australia, Brazil, Sweden, Japan, Singapore, Malaysia, Vietnam, China Hong Kong, China Macao, and China Taiwan.

6.1.3.1 PROMS 2021 Keynote Speeches

The theme of PROMS2021 is "The Application of Rasch Model in Psychological and Educational Research". The organizer invited four distinguished keynote speakers. The keynote speech by Professor Trevor Bond from James Cook University in Australia is titled "From Estimation to Consideration: The Role of Rasch Model Measurement in Promoting Understanding and Enhancing Validity"; The keynote speech by Professor Kit Tai HAU from The Chinese University of Hong Kong is titled "Large Scale International Education Assessment: Uses, Limitations, and Counterintuitive Discoveries"; The keynote speech by Professor Ricardo Primi from São Francisco University in Brazil is titled "Response Style as Individual Differential Function: Multiple Methods for Addressing Individual DIF"; Professor Steven Stemler from Wesleyan University in the United States gave a keynote speech titled "Less Parameters, Better Results! The True Value of Rasch Model over Project Response Theory".

In addition, the conference also invited Professor Quan Zhang from Jiaxing University in Zhejiang, China to give a special report entitled "Rasch Model and Test Equating: The Most Updated Rasch research in China: RASCH-GZ". Professor Zhang demonstrated online the entire process of parallel test equating to the participants. He emphasized that the RASCH-GZ system is a fully upgraded system based on GITEST for both item analysis and test equating, and it is the only one Rasch-based software so far successfully developed across China. This development shows that Chinese scholars have occupied a place in the international Rasch model research field and also represents the development process and trend of Rasch model research in China.

6.1.3.2 PROMS2021 Pre-conference Workshops

The organizers of PROMS2021 arranged one-day pre-conference workshops, which were divided into two parallel themes.

Theme (1) is the "Rasch Model Workshop" hosted by Dr. Zi Yan, associate professor of Curriculum and Teaching at the Education University of Hong Kong and current vice chairman of PROMS. Dr. Zi Yan emphasized that the Rasch model focuses on items and respondents rather than test scores by integrating the two through the use of joint measurement principles; the quantitative analysis of qualitative problems is reflected in a way that is rarely seen in social sciences. The workshop also introduced the advantages of analyzing data based on the Rasch model compared to classical testing and questionnaire scoring methods, such as how data analysis based on the Rasch model can be applied to secondary scoring data (basic Rasch model) and Likert questionnaire data (Rasch model scoring scale model).

The workshop theme (2) was run by two experts. Dr. Cynthia Tong from the Department of Psychology at the University of Virginia in the United States hosted the "R Language Workshop", while Dr. Kit Tai HAU, professor from the Department of Educational Psychology at the Chinese University of Hong Kong, China, hosted the "Educational Data: Storytelling and Methods for Finding Teaching Implications" workshop. The "R Language Workshop" mainly introduces "What is R" and its application in Rasch model measurement. This workshop conducted in-depth discussions on topics such as t-tests, analysis of variance, multiple regression, confirmatory factor analysis (CFA), measurement invariance, and structural equation modeling (SEM). The workshop on "Education Data: Storytelling and Methods for Finding Teaching Implications" covers: (1) How to select questionnaires for educational monitoring; (2) How to use questionnaire results to analyze and write attractive and influential "stories". The workshop uses PISA-type tools and their materials as case studies and provides explanations. This workshop is dedicated to creating vivid and interesting "stories" for training, rather than conducting complex and advanced statistical techniques.

6.1.4 The First PROMS On-Site Conference after the COVID-19 Pandemic Period

In 2023, after the epidemic, we open our arms to embrace the first on-site conference of PROMS2023 in Macau, SAR, China. This is a grand international conference after three years of global fighting epidemic efforts. We did have careful consideration and finally decided to have the conference venue in Macau, China. Why is it in Macau? For ordinary people, Macau is a paradise for gamblers. But we have our own unique insights. For most people's understanding, Macau is a paradise for casinos, but for our Rasch team, Macau can also be said to be the birthplace of modern probability and statistics. The concepts of probability and statistics play an important role in Macau society. Without casinos, there would be no development of modern probability and statistics. Without probability and statistics, scientific research cannot be very well conducted, including the research and application of Rasch models. Therefore, holding an academic conference like PROMS in Macau is the best choice. Apart from this, Macau is a multicultural and fully open city, making it very convenient for participants from various countries and regions to travel. Of course, there is also Macau's cuisine for the celebration after the COVID-19 pandemic period. Therefore, PROMS2023 was successfully held on 29th–30th August 2023 at the World Sports University (WSU), Macau, China.

This is the true international conference successfully held in China after the global fighting against the COVID-19 epidemic years since 2019. A truly international conference attracting experts, researchers and scholars from the United States, Australia, Indonesia, Japan, Malaysia, Singapore, South Korea, the Philippines, China Hong Kong, China Macau, and China Mainland. ROMS2023, Macau, China is the premier international venue for Rasch researchers and practitioners to share new ideas, research results, and development experiences. Apart from keynotes and parallel sessions, three parallel pre-conference workshops were conducted.

The theme of PROMS2023 is "Rasch Model: Tradition, Challenges, and Trends after COVID-19". As we continue to navigate the post-pandemic landscape, we are faced with new challenges and opportunities to advance our understanding of objective measurement in various fields. To quote Dr. Che Kuong Hon, the president of WSU in his welcome speech, at the WSU, a global institution that is committed to advancing knowledge and research in the field of sports, physical education, and related disciplines, we believe that objective measurement is essential for advancing knowledge and understanding in our field. The WSU strives to promote excellence in teaching, research, and service, and to foster collaboration across cultures and borders. Our mission is to create a vibrant academic community that is dedicated to improving the lives of people through sports and physical activity.

6.1.5 PROMS Tenet

The principle aims of PROMS (the society) are to:

- Encourage the pursuit of Rasch measurement in the Pacific region and its application in the fields of Business, Counselling, Economics, Education, Health Care, Language, Measurement, Psychology, Quality Assurance, Statistics, and Strategic Planning.
- Advocate measurement practice that contributes to individual, community, and societal well-being, and the public good, in the Pacific region.
- Raise the visibility and status of Rasch measurement in the Pacific region.

These aims are to be accomplished through:

- Establishing a Pacific region network of researchers and scholars in order to foster cooperative and collaborative relationships.
- Holding an annual meeting—the PROMS
- Providing workshops and training.
- Engaging in such other activities as might facilitate the development of research, policy, and practice in the Rasch measurement.

The three goals of PROMS inspire, promote and increase the application of measurement based on the Rasch model in both industries and academic sectors that contribute to the development in the Pacific Rim and beyond.

6.1.6 An Overview of PROMS Conferences over the Past Two Decades (2005–2025)

Table 6.1 presents the PROMS Conferences held in different countries and regions over the past two decades.

TABLE 6.1

A List of the PROMS Conferences Held in Different Countries and Regions over the Past Two Decades.

Countries/ Regions	Years	Organizers	Venues
Malaysia	The 1st PROMS 2005 June, 2005	Hosted by Prof Noorlide Abu Kasim and Prof. Nor Akbariah Makhzir	International Islamic University of Malaysia (IIUM) Kuala Lumpur, Malaysia
	The 6th PROMS 2010 Dates: 29 June to 1 July 2010	International Islamic University of Malaysia, Kuala Lumpur, Malaysia	International Islamic University of Malaysia, Kuala Lumpur, Malaysia
	The 13th PROMS 2017 Dates: 5–9, August, 2017	Universiti Malaysia Sabah, Malaysia	Kota Kinabalu, Malaysia Universiti Malaysia Sabah, Malaysia
	The 19th PROMS 2024 Dates: 19–21 August 2024	Hosted by Prof. Harris Shah Abd Hamid Malaysia University College MAIWP International	Malaysia University College MAIWP International Kuala Lumpur, Malaysia
China Hong Kong	The 2nd PROMS 2006	Hosted by Prof. Magdalena Mo Ching Mok The Hong Kong Institute of Education, Hong Kong, SAR, China	The Hong Kong Institute of Education, Hong Kong, SAR, China
	The 5th PROMS 2009 Dates: 28–30 July 2009	The Hong Kong Institute of Education, Hong Kong, SAR, China	The Hong Kong Institute of Education, Hong Kong, SAR, China
China Macau	The 18th PROMS2023 Dates: 28–30 August 2023	Hosted by Prof. Quan Zhang The World Sports University Macau, SAR, China	The World Sports University Macau, SAR, China
China Taiwan	The 3rd PROMS 2007 Dates: 5–7 August 2007	National Chung Cheng University	National Chung Cheng University, Chiayi, Taiwan, China
	The 9th PROMS 2013 Dates: 3–5 August 2013	National Sun Yat-sen University, Kaohsiung, China	National Sun Yat-sen University, Kaohsiung, China

(Continued)

TABLE 6.1 (*Continued*)

A List of the PROMS Conferences Held in Different Countries and Regions over the Past Two Decades.

Countries/ Regions	Years	Organizers	Venues
China Mainland	The 8th PROMS 2012 Dates: 6–9 August 2021	Hosted by Prof. Quan Zhang Jiaxing University, Zhejiang, China	Jiaxing University, Zhejiang, China
	The 10th PROMS 2014 Dates: 4–6 August 2014	Hosted by Prof. Quan Zhang Jiaxing University, Zhejiang, China	Clifford Hotel Panyu, Guangzhou, China
	The 12th PROMS 2016 Dates: 30–31 July 2016	Hosted by Prof. Quan Zhang Jiaxing University, Zhejiang, China	Oriental Hotel Xi'an, Shaanxi, China
	The 14th PROMS 2018 Dates: 23–27 July 2018	Hosted by Dr. Jason Fan College of Foreign Studies Fudan University Shanghai, China	Fudan University Shanghai, China
	The 16th PROMS 2021 Dates: 4–6 December 2021	Hosted by Prof. Chen Chen Department of Psychology Nanjing Normal University, Nanjing, China	Online ZOOM
Japan	The 4th PROMS 2008 Dates: 31 July to 3 August 2008	Ochanomizu University Tokyo, Japan	Ochanomizu University Tokyo, Japan
	The 11th PROMS 2015 Dates: 20–24 August 2015	Kyushu Sangyo University Fukuoka, Japan	Kyushu Sangyo University Fukuoka, Japan
Singapore	The 7th PROMS 2011 Dates: 13–15 July 2011	National Institute of Education, Singapore	National Institute of Education, Singapore
	The 20th PROMS2025 Dates: 21–23 July 2025	Hosted by Dr. Lye Che Yee Singapore University of Social Sciences (SUSS) Singapore	Singapore University of Social Sciences (SUSS) Singapore
Indonesia	The 15th PROMS 2019 Dates: 26–31 August 2019	Hosted by Prof. Bambang Sumintono and Indonesian Association of Psychometrics (Apsimetri) & Faculty of Psychology, Universitas Airlangga, Surabaya, Indonesia	Bumi Surabaya City Resort, Surabaya, Indonesia

TABLE 6.1 (***Continued***)

A List of the PROMS Conferences Held in Different Countries and Regions over the Past Two Decades.

Countries/ Regions	Years	Organizers	Venues
Vietnam	The 17th PROMS 2022 Dates: 2–3 December 2022	Hosted by Dr. Đặng Xuân Cương Research Division on Educational Assessment The Vietnam National Institute of Educational Sciences (VNIES), Ministry of Education and Training, Hanoi, Vietnam	Online ZOOM

6.2 Features of PROMS Pre-Conference Workshops

Based on the practical experience, the most effective way to promote the learning and application of Rasch model is to run pre-conference workshops. The PROMS pre-conference workshops can be described as: International, Professional, and Pragmatic. At the first place, for over two consecutive decades, the contents remain unchanged. Just focusing on promoting the application of Rasch model and the introduction to and the specific use of Rasch-based software (see Section 3.3). All the workshop runners are well-known professionals in the Rasch research field coming from Australia, USA, UK, China, Japan, Malaysia, Singapore, and China Hong Kong. This includes Trevor Bond, Mike Linacre, David Andrich, Jack Stenner, George Engelhard, Haniza, etc. The workshop usually runs from one to two days, falling into three phases. The first phase focuses mainly on the promoting of the theory and application about the Rasch measurement followed by the demonstration, the use and the training of the specific Rasch-based software like WINSTEPS, FACET, and so on. Due to the practicality of the workshop, participants needn't have the particular background of psychometric competence or psychological measurement. They only bring their laptops, download and install the free software link with password provided by workshop instructors on site. Once installed successfully, the software is available to use. The most important phase of these pre-conference workshops is the demonstration and question & answers session, which has long been taken as popular classes for international participants to learn Rasch model for conducting research as well as data analysis. On occasions when many participants are non-English speakers, interpretation in local language is provided to help understand the instructions given by the workshop runners. Over the past two decades, the popular workshop topics among learners are listed below:

6.2.1 Winsteps and Facets

The first pre-conference workshop conducted by Prof. Mike Linacre at PROMS2005, Kuala Lumpur, Malaysia. This workshop laid the foundation and direction for future workshops as many first-time delegates from Southeast Asia were not familiar with WINSTEPS and FACET at that time.

6.2.2 Introducing Rasch Measurement Using WINSTEPS

This was jointly conducted by Prof. Trevor Bond from James Cook University and Prof. Zi Yan of the Education University of Hong Kong SAR, China. This practical workshop was conducted in both English and Chinese to introduce Rasch measurement to colleagues using Winsteps software. Theoretical and procedural presentations were followed by guided hands-on data analysis and interpretation. Participants were told to bring their own laptop computers (with Office and Adobe Reader pre-installed) to the workshop.

6.2.3 The Generalized Linear Mixed Models and the Rasch Measurement Theory

This was jointly conducted by Prof. George Engelhard, Jr. from the University of Georgia, USA and Prof. Jue Wang from University of Science and Technology of China. The purpose of this workshop is to introduce participants to the use of generalized linear mixed models and Rasch Measurement Theory. An overview of generalized linear mixed models is provided. The two professors illustrated the use of an R package called EIRM (Bulut, 2021) to estimate several extended Rasch models, including the Dichotomous Model, the Linear Logistic Rasch Model, Linear Regression Rasch Model, and Combined Covariates Rasch Model.

6.2.4 The Applications of the Lexile Framework for Reading

This was run by Prof. Jack Stenner, the president of the Institute for Objective Measurement, USA.

The full-day workshop provides extensive interactive training on how the Lexie Framework for Reading was built. Attendees will participate in group activities that provide hands-on learning experiences that can impact their day-to-day assessment and psychometric practice. Topics in Jack's workshop covered include:

- Development of The Lexile Framework for Reading;
- Use of a Causal Rasch model;
- Testing construct theories;
- The Reading Task Continuum: Adjusting measures for added easiness—hardness and differences in unit size;

- Introduction to EdSphere[2]—a personalized learning platform for practicing reading (all participants receive a free one-year license to EdSphere);
- Understanding Lexile measures as indicators of text complexity and reading level;
- Using Lexile measures to match readers to text and forecast comprehension;
- Track student progress toward college and career readiness: Growth Models;
- Understand the relationship between Lexile measures and the Common Core State Standards;
- A guided tour of www.Lexile.com

6.2.5 TAM (Test Analysis Modules)

This was run by Dr. Margaret Wu from Victoria University, Melbourne, Australia. This is a two-day workshop on the use of an IRT software TAM. TAM is written in R and is free for download. TAM fits the Rasch model and 2PL models. It can estimate unidimensional and multidimensional models with latent regression and facet terms. Both joint maximum likelihood (JML) and marginal maximum likelihood (MML) estimation methods can be used. TAM provides functions for computing fit statistics and drawing plausible values. It is fast to run. Participants are requested to bring their personal computers with R studio installed. R studio is free and can be downloaded from the internet. Instructions for downloading TAM are provided before the workshop.

6.2.6 The BEAR ASSESSMENT System, Large Scale Assessment, and Explanatory Measurement

This was conducted by Prof. Mark Ronald Wilson from the University of California, Berkeley, USA. This workshop is divided into three parts. In Part I, the instructor introduces the BEAR Assessment System (BAS). The BAS was designed to build upon methodological and conceptual advances in assessment. The instructor outlines the four principles and the four "building blocks" of the assessment system, and provides an example of each of these components from work on a learning progression in statistics and modeling for middle schools. In Part II, the instructor explores implications of the BAS for large-scale assessment. The argument here is that the current emphasis on testing in education has led to a "squeeze" that teachers and their students experience between the need to cover many standards, and the educator's wish to value meaningful instruction and assessment. This is illustrated in the context of the same example as in Part 1. In Part III, the instructor describes a role for the BAS in explanatory measurement. Explanatory measurement emphasizes how item response models can be coordinated and broadened to emphasize their explanatory uses beyond their standard descriptive uses. The basic explanatory principle is that item responses can be modelled as a function of qualities and features of various measurement contexts. According to the workshop instructor, these qualities and features can be: (a) characteristics of (i) items, of (ii) persons, and of (iii) combinations of items and persons; (b) observed or latent (of either items or persons); and (c) continuous or categorical. These ideas are exemplified in the context of a reading comprehension test.

6.2.7 Item Analysis and Test Equating

This was jointly conducted (since 2021) by Prof. Quan Zhang from Jiaxing University, China/The World Sports University, Macau, China, and his student Jingang Wei who is a computer engineer on Program of Education, Tarlac State University, the Philippines. This workshop focuses on two important aspects: CTT-based item analysis and Rasch-based Test equating. The greatest advantage of Rasch GZ is that users do not need to edit complex command files. As long as the data files are prepared according to the format requirements, they can complete the operations according to the menu. At the same time, the latest version of Rasch GZ combines modern AI technology to provide voice or screen text prompts for any errors that occur during the data file editing process to ensure correct operation. In terms of output files, users can choose the language they prefer and print the output files in PDF or WORD documents. Table 6.2 provides the detailed list of PROMS Pre-Conference workshops over the past two decades.

TABLE 6.2

List of Pre-Conference Workshops at PROMS International Conferences over the Past Two Decades

Years and Venues	Topics	Workshop Runners	From Universities
The 1st PROMS 2005, Malaysia Dates: June 2005 Venue: International Islamic University of Malaysia (IIUM) Kuala Lumpur, Malaysia Hosted by Prof Noorlide Abu Kasim and Prof. Nor Akbariah Makhzir	Winsteps and Facets	Prof Mike Linacre	A prominent figure in the field of psychometrics, particularly known for his work with Rasch measurement theory. Mike has made significant contributions to the development and application of Rasch models in various fields, including education, psychology, and health sciences.

TABLE 6.2 (*Continued*)

List of Pre-Conference Workshops at PROMS International Conferences over the Past Two Decades

Years and Venues	Topics	Workshop Runners	From Universities
The 2nd PROMS 2006, China Venue: The Hong Kong Institute of Education Hong Kong, SAR, China	Two workshops aimed at providing participants with in-depth knowledge and skills related to measurement and assessment techniques.	Prof. Trevor Bond Ray Adams[3]	James Cook University, Australia ACER
The 3rd PROMS 2007, Taiwan, China Venue: Taiwan, China	Three workshops[4] were designed to provide attendees with specialized training and insights on topics pertinent to the conference's focus.	Trevor Bond Ray Adams Jue Wang	James Cook University, Australia ACER Hong Kong Institute of Education, Hong Kong, China
The 4th PROMS2008, Japan Dates: 31 July–3 August 2008 Venue: Ochanomizu University, Tokyo, Japan	Theme: Objective Measurement for Teaching, Learning, and Living Two pre-conference workshops[5] typically aimed to offer participants specialized training and insights into topics relevant to the conference's themes. For more detailed information about the content and focus of these workshops, you may want to refer to conference archives or official materials from the event organizers.	Trevor Bond Ray Adams Jue Wang	James Cook University, Australia University of Science and Technology of China, China
The 5th PROMS 2009, Hong Kong, SAR, China Dates: 28–30 July 2009 Venue: The Hong Kong Institute of Education, Hong Kong, China	Theme: Quality Rasch Measurement Three pre-conference workshops[6] were designed to provide participants with in-depth knowledge and hands-on experience in specific areas related to the conference's focus.	Trevor Bond Jue Wang	

(*Continued*)

TABLE 6.2 ***(Continued)***

List of Pre-Conference Workshops at PROMS International Conferences over the Past Two Decades

Years and Venues	Topics	Workshop Runners	From Universities
The 6th PROMS 2010, Malaysia Dates: 29 June–1 July 2010 Venue: International Islamic University Malaysia (IIUM), Kuala Lumpur, Malaysia	Four workshops were designed to provide participants with in-depth knowledge and hands-on experience in specific areas related to the conference's focus.	Ray Adams	ACER
		Geoff Masters	ACER
		Jue Wang	The Hong Kong Institute of Education, Hong Kong, China
		C.Y. Tang	National Taiwan Normal University, China
The 7th PROMS 2011, Singapore Dates: 13–15 July 2011 Venue: National Institute of Education, Singapore	Objective Measurement for Teaching, Learning, and Living Three workshops were designed to provide participants with in-depth knowledge and hands-on experience in specific areas related to the conference's focus.	Trevor Bond	James Cook University, Australia
		Ray Adams	From ACER
		Jue Wang	
The 8th PROMS 2012, China Dates: 6–9 August Venue: Jiaxing University, China The first time in China Mainland	Introduction to the Rasch Measurement via WINSTEPS	Prof. Trevor G. Bond Prof. Zi Yan	The Education University of Hong Kong, SAR, China
	Introduction to Many-Facets Rasch Measurement Using FACTES	Dr. James Sick	International Christian University, Tokyo Japan
	EQSIRT	Dr. Eric Wu	Department of Psychology, University of California, Los Angeles (UCLA), USA
	Measuring English Language Performance with Lexiles	Prof. Jackson Stenner	Chairman and CEO and Co-founder Meta Metrics Inc., USA President of the Board of Directors of Institute of Objective Measurement, a board member for the National Institute for Statistical Sciences, and a past board member for Duke Children's Hospital and the North Carolina Electrics and Information Technologies Association, USA

TABLE 6.2 (*Continued*)

List of Pre-Conference Workshops at PROMS International Conferences over the Past Two Decades

Years and Venues	Topics	Workshop Runners	From Universities
The 9th PROMS2013, China Dates: 3–5 August 2013 Venue: National Sun Yat-sen University, Kaohsiung, China	Measurement must balance its statistical and scientific elements	Prof. Mark Ronald Wilson	University of California, Berkeley, USA
	Research on and Application of The Lexile Framework for Reading	Prof. Jack Stenner	Chairman and CEO and Co-founder Meta Metrics Inc., USA President of the Board of Directors of Institute of Objective Measurement, a board member for the National Institute for Statistical Sciences, and a past board member for Duke Children's Hospital and the North Carolina Electrics and Information Technologies Association, USA
	TAM (Test Analysis Modules)	Dr. Margaret Wu	Victoria University, Melbourne, Australia
The 10th PROMS 2014, China Dates: 4–6 August 2014 Venue: Clifford Hotel, Panyu, Guangzhou, China Hosted by Prof Quan Zhang, representing Jiaxing University, China	Rasch Measurement Using WINSTEPS	Prof. Trevor G. Bond Prof. Zi Yan	James Cook University, Australia The Education University of Hong Kong, SAR, China
	Introduction to and Demonstration of Using TAM Software for IRT Analysis	Prof. Margaret Wu	Victoria University Melbourne, Australia
	Invariant Measurement with Raters and Rating Scales	Prof. Jackson Stenner	Chairman and CEO and Co-founder Meta Metrics Inc., USA President of the Board of Directors of Institute of Objective Measurement, a board member for the National Institute for Statistical Sciences, and a past board member for Duke Children's Hospital and the North Carolina Electrics and Information Technologies Association, USA
	Introduction to Lexiles and Developing Construct Models	Prof. George Engelhard Jr.	The University of George, USA

(*Continued*)

TABLE 6.2 (*Continued*)

List of Pre-Conference Workshops at PROMS International Conferences over the Past Two Decades

Years and Venues	Topics	Workshop Runners	From Universities
The 11th PROMS2015, Japan Dates: 20-24 August 2015 Venue: Kyushu Sangyo University Fukuoka, Japan	Item Banking Using the Rasch Measurement Model	Dr. Lead Dr. Sassoil Sadeghi	Australia Curriculum, Assessment and Reporting Authority (ACARA)
	An Introduction to Rasch Analysis Using WINSTEPS	Prof. Trevor G. Bond	James Cook University, Australia
	The Concept of Causal Models	Prof. Jack Stenner	Chairman and CEO and Co-founder Meta Metrics Inc., USA, President of the Board of Directors of Institute of Objective Measurement, Board member for the National Institute for Statistical Sciences, and a past board member for Duke Children's Hospital and the North Carolina Electrics and Information Technologies Association, USA
	Introduction to Many Facet Rasch Model	Dr. James Sick	International Christian University, Tokyo
	Basic Concepts of Computer-Adaptive Testing (CAT) and an Opportunity to Implement a Rasch-based Small Scale CAT	Prof. Tetsuo Kimura	Niigata Seiryo University, Japan
	Invariant Measurement with Raters and Rating Scales	Prof. George Engelhard Jr.	The University of George, USA
The 12th PROMS 2016 Xi'an, China Dates: 30-31 July 2016 Venue: Oriental Hotel Xi'an, China Hosted by Prof Quan Zhang, representing Jiaxing University, China	Introduction to Rasch Measurement Using WINSTEPS	Prof. Trevor G. Bond Prof. Zi Yan	James Cook University, Australia The Education University of Hong Kong, SAR, China

TABLE 6.2 (*Continued*)

List of Pre-Conference Workshops at PROMS International Conferences over the Past Two Decades

Years and Venues	Topics	Workshop Runners	From Universities
	An Introduction to the Analysis of Data	Prof. John Barnard	IACAT
	Substantive Theory General Objectivity and an Individual-Centered Psychometrics	Prof. Jack Stenner	Chairman and CEO and Co-founder Meta Metrics Inc., USA President of the Board of Directors of Institute of Objective Measurement, a board member for the National Institute for Statistical Sciences, and a past board member for Duke Children's Hospital and the North Carolina Electrics and Information Technologies Association, USA
	Application of Many-Facet Rasch Model in Language Assessment: An Introduction	Dr. Jason Fan	Fudan University, Shanghai, China
The 13th PROMS 2017, Kota Kinabalu, Malaysia Dates: 5–9 August 2017 Venue: Universiti Malaysia Sabah	Introduction to Rasch Measurement Model	Prof. Trevor Bond & Dr. Mohd Zali	James Cook University, Australia myRasch Malaysia
	Introduction to SEM and Rasch Measures	Dr. Juliet Ling Mei Teng Dr. Nor Irvoni Mohd Ishar	Universiti Malaysia Sabah Universiti Teknologi MARA, Malaysia
The 14th PROMS 2018, Shanghai, China Dates: 23–27 July 2018 Venue: Fudan University, China Hosted by Fudan University, China	Using Rasch Model in Language Assessment: An introduction to WINSTEPS	Prof. Trevor G. Bond Prof. Zi Yan	James Cook University, Australia The Education University of Hong Kong, SAR, China
	Using Rasch Model in L2 Writing Research: An introduction to FACETS	Dr. Xun Yan Dr. Jason Fan	University of Illinois at Urbana-Champaign, USA Fudan University, China
	Applying Extended Rasch Models Using R and R-Studio	Dr. Hudson Golino	The University of Virginia, USA

(*Continued*)

TABLE 6.2 (*Continued*)

List of Pre-Conference Workshops at PROMS International Conferences over the Past Two Decades

Years and Venues	Topics	Workshop Runners	From Universities
The 15th PROMS2019, Surabaya, Indonesia Dates: 26–31 August 2019 Venue: Bumi Surabaya City Resort	Using R: From CTT to IRT	Urip Purwono., M.S., M.Sc., Ph.D. Sukaesi Marianti., S.Psi., M.Si., PhD	Universitas Padjadjaran, Bandung, Indonesia Universitas Brawijaya, Malang, Indonesia
Hosted by Indonesian Association of Psychometrics (Apsimetri) & Faculty of Psychology, Universitas Airlangga, Surabaya, Indonesia	Introduction of the Rasch Model for Measurement and Research in Social Science	Prof. Yan Zi	The Education University of Hong Kong, SAR, China
	Rasch Modeling for Quantitative Research	Prof. Bambang Sumintono Ph.D.	Universitas Malaya, Kuala Lumpur, Malaysia
	Test Scoring: Obtaining True Score When the Assumption of Unidimensionality is not Tenable	Jahya Umar., Ph.D.	Universitas Islam Negeri Syarif Hidayatullah, Jakarta, Indonesia
	Computer Programing for Psychophysical Measurement	Galang Lufityanto., M.Psi., Ph.D.	Universitas Gadjah Mada, Yogyakarta, Indonesia
	Ethical Guidelines and Professional Standards in Psychological Measurement	Dr. Ide Bagus Siaputra., S.Psi	Universitas Surabaya, Surabaya, Indonesia
The 16th PROMS 2021, Nanjing, China Dates: 4–6 December 2021 Venue: Nanjing Normal University, China	Generalized Linear Mixed Models and Rasch Measurement Theory	Prof. George Engelhard Jr. Prof. Jue Wang	The University of George, USA University of Science and Technology of China, China
The first large-scale online PROMS during the global fighting against COVID-19 pandemic	Introducing Rasch Measurement Using WINSTEPS.	Prof. Zi Yan	The Education University of Hong Kong SAR, China
	Rasch-GZ: The fully updated (GITEST) Version	Prof. Quan Zhang	Jiaxing University, China
The 17th PROMS 2022, Hanoi, Vietnam Dates: 2–3 December 2022 Venue: The Vietnam National Institute of Educational Sciences, Hanoi, Vietnam	Online PROMS during the global fighting against COVID-19 pandemic		

TABLE 6.2 (*Continued*)

List of Pre-Conference Workshops at PROMS International Conferences over the Past Two Decades

Years and Venues	Topics	Workshop Runners	From Universities
The 18th PROMS2023, Macau, China Dates: 29–30 August 2023 Venue: The World Sports University, Macau, SAR, China	Generalized Linear Mixed Models and Rasch Measurement Theory	Prof. George Engelhard Jr. Prof. Jue Wang	The University of George, USA University of Science and Technology of China, China
	Introduction to Rasch Measurement Using WINSTEPS	Prof. Zi Yan	The Education University of Hong Kong
	Introduction to Rasch_GZ: The 1st Chinese Version of Rasch-based Item Analysis and Test Equating	Prof. Quan Zhang Dr Jingang Wei	Jiaxing University, China Computer Engineer, City Training Institute (CTI), Guangzhou, China MA-PhD Candidate of Program of Education, Tarlac State University, the Philippines
The 19th PROMS2024, Kuala Lumpur, Malaysia Dates: 19–21 August 2024 Venue:	Introduction to Rasch Measurement Model	Prof. Trevor G. Bond, Dr. Mohd Zali	James Cook University, Australia International Islamic University of Malaysia (IIUM) Newstar Agencies S/B, Malaysia
	Rasch-GZ: The 1st Chinese Version of Rasch-based Item Analysis and Test Equating	Prof. Quan Zhang Dr Jingang WeiJ	Jiaxing University, China City Training Institute (CTI), Guangzhou, China
The 20th PROMS2025, Singapore Dates: 21–23 July 2025 Venue: Singapore University of Social Sciences (SUSS)	Adapting Rasch Meta-meters of Growth for Variables in the Social Sciences	Emeritus Professor David Andrich,	The University of Western Australia
	Rasch Measurement for Rater-mediated Assessment	Prof. George Engelhard Jr. Prof. Jue Wang	The University of George, USA University of Science and Technology of China
	Rasch-GZ: Item Analysis and Test Equating in the Age of AI	Prof. Quan Zhang Dr Jingang Wei	Jiaxing University, China City Training Institute (CTI), Guangzhou, China
	Item Banking and Adapting Testing	Dr. Sassoul Sadeghi	Australian Curriculum, Assessment and Reporting Authority

6.3 Keynote Topics at PROMS Conferences over the Past Two Decades (2005–2025)

According to the incomplete statistics, from 2005 to 2025, approximately 80 well-known keynote speakers were invited from universities in Australia, Brazil, Malaysia, Denmark, Finland, Indonesia, Japan, Singapore, Sweden, the United States, Vietnam, China Hong Kong, China Macau, China Taiwan, and China Mainland. In particular, PROMS conferences held over the past 20 years have tested the qualification of Rasch scholars in the research and application of the Rasch measurement theory.

The content and scope of PROMS keynote speeches (may include but are not limited to The Rasch model) can be roughly divided into three stages. In the first decade, the focus was mainly on "Rasch Model: Theory, Method and Practice", i.e., Rasch Measurement and Scales, then with discussions on other Rasch-related topics mostly including technical issues of the application of Rasch model, measurement across cultures and fairness of using assessment to make decisions, assessing creativity and other models related to Rasch models. In recent years, with the rise of artificial intelligence (AI), the topics of keynote speeches have also kept pace with the times, incorporating AI content into the research and application of Rasch models.

6.3.1 Rasch Measurement and Scales

The first keynote speech addressed by Prof. Trevor Bond at the first PROMS2005, Kuala Lumpur, Malaysia, is "Achieving the impossible: Solving real measurement problems with the Rasch model", which lays the solid foundation for the Rasch measurement research and development in the following years.

In PROMS2024 held at University College MAIWP International, Kuala Lumpur, Malaysia, Prof. Harris Shah Abd Hamid,[7] the host was invited to deliver a keynote speech titled Mapping the Rating Mind: Comparison of Rating Categories in Personality and Motivation Measures. In his speech, Harris addressed that using a rating scale (e.g., 1 = strongly disagree, 5 = strongly agree) is a convenient way to quantify latent psychological constructs. However, the interpretation of scores generated from such scales should not be taken for granted. An experimental approach, grounded in objective measurement, involves examination of the rating scales' functionality and modification, where necessary, before meanings are assigned to the scores. Harris had a big sample size. A comparison of seven ASEAN countries revealed differences of average scores for categories used in rating the justifiability of social behaviors. Furthermore, the distributions of frequency of rating categories usages are not uniform and deserve more detailed examination. A rationale for such an examination lies in the differences of rating categories usage and Rasch scores. Harris also used calibrated data for a set of psychological inventory to compare the rating categories' functioning for personality traits (conscientiousness, openness, neuroticism, and extraversion) and motivation (need for power, need for achievement, and need for affiliation) measures. Harris concluded that these findings give rise to a postulate that rating behaviors provide insights into how people react cognitively to items measuring different psychological constructs.

6.3.2 Testing, Equating, Fairness, and Learning

In PROMS2012, for the first time, Professor Quan Zhang from Jiaxing University delivered the keynote speech to report to the Rasch community how they successfully completed the Ten-Year Matriculation English Test (MET) Equating Project via the Rasch model under the guidance of the late famous Chinese linguist Professor Shichun Gui (1930–2017). The equating project was financially sponsored by the Examination Center under the Ministry of Education, China (1990–1999). This is the first time that MET Equating Project on China national level has been revealed to the international counterparts in English.

In PROMS2013 organized by National Sun Yat-sen University, Kaohsiung, China, Dr. Hua-Hua Chang from University of Illinois at Urbana-Champaign, USA presented keynote on "Combining Adaptive Testing with Adaptive Learning". Prof. Chang said that computerized adaptive testing (CAT) has become fashionable in large-scale high-stake assessments. A major benefit of using CAT is that it measures the latent trait level of examinees with greater precision than conventional non-adaptive tests by building an individualized test for each examinee. In Chang's speech, the prospect of using CAT to facilitate "adaptive learning" is introduced. Note that tailoring instruction according to each student's ability level has always been our dream. With the help of CAT, it is possible to make classroom instruction individualized.

In PROMS2014, organized jointly by Jiaxing University and City Training Institute, (CTI) Guangzhou, China, Prof. Yan Jin[8] from Shanghai Jiaotong University and Dr. Eric Wu[9] from UCLA were invited to jointly deliver the keynote speech: "An Argument Approach to Test Fairness: The case of multiple-form equating in the College English Tests (CETs)".

By citing many experts of language testing, Jin and Wu addressed that test fairness, in the view of some researchers (e.g., Willingham Cole, 1997; Xi, 2010), is conceptualized as an aspect of validity, that is, a test has to be fair to be valid. Whatever weakens fairness also compromises the validity of a test. In comparison with the continued development of conceptual frameworks for test fairness (e.g., Kunnan, 2000, 2004), few empirical studies, however, have been conducted to tackle fairness-related problems in language testing. To address the lag of empirical fairness research in language testing, the two speakers cited Xi (2010) who proposed an argument approach to examining test fairness, which conceptualizes fairness as comparable validity for identifiable and relevant groups across all stages of assessment.

In this keynote, the argument approach is adopted to examine, from the test developer's perspective, a rebuttal of score equating that would weaken the fairness argument of the CET in China, a test of very high stakes and an ultra-large scale (see Jin, 2010; Yang, 2003; Zheng Cheng, 2008). Following an introduction to the multiple-form measure recently implemented to prevent high-tech cheating in the CET, they discuss the design of multiple-form equating to adjust the differences in difficulty across alternate forms of one test so as to produce comparable score scales. Based on random samples on each test form, results of the traditional equi-percentile equating and the IRT true score equating (Harris Kolen, 1986; Kolen, 1988) are compared in terms of their equating accuracy, efficiency, and practicality.

To ensure test fairness, it is important that construct-irrelevant factors such as variation in test item/form difficulty produce no systematic and appreciable effects on test results. The investigation of the procedures for test form equating represents the CET developers' efforts to conduct empirical fairness research in a principled way. This keynote speech has shown that the argument approach, though a topic still being hotly debated, "can be particularly useful in focusing research on specific threats to fairness/validity" (Kane, 2010, p. 177).

Again, in PROMS 2014, Guangzhou, China, a research team of language test developers headed by Dr. Alan Urmston[10] from Hong Kong Polytechnic University with his two members Dr. Michelle Raquel[11] and Dr. Gwendoline Guan[12] was invited to jointly deliver a keynote speech about DELTA: A system for diagnosing and tracking English competence in Chinese university students. This is also a much earlier computer-based online assessment system ever developed based on Rasch measurement principles with big sample size.

The team reported on the development of a post-entry diagnostic English language assessment for undergraduate students studying at Hong Kong universities. Three universities provide English language support to their mostly Chinese students throughout their studies through the implementation of the DELTA: Diagnostic English Language Tracking Assessment. DELTA, a computer-based online assessment system, based on Rasch measurement principles, was created to enable students to monitor their English language proficiency throughout their degree programs. DELTA is able to diagnose and track students' progress in enhancing their English language proficiency while at university. The presenters discussed how DELTA operates within the context of Hong Kong, how Rasch measurement is used to support diagnosis and tracking functions, as well as current work to develop the listening component of the system. The keynote was chaired by Prof. Bond, and in the Q & A session, the issues involved in diagnosing and tracking English language proficiency for students whose first language is Cantonese or Mandarin were fully discussed.

In PROMS 2016 held in Xi'an, China, Professor Xiaoting Huang[13] of Peking University was invited to deliver a keynote speech titled: "Investigating the Predictive Validity and Social Consequences" of "Gao-kao[14]". According to Prof. Huang, the Chinese College Entrance Examination ("Gao-Kao") is the highest-stakes assessment in China and parallels the most competitive examinations globally. As universities base their admission decisions solely on students' test scores up till now, the test's reliability and validity can never be overemphasized. Although China's NEEA (National Education Examinations Authority) exerts great effort to ensure the quality of the test items and the reliability of the test, empirical studies on its consequence validity are scarce due to the lack of data. In Huang's speech, she analyzed University K's 2004–2009 cohort data to examine the validity of Gao-Kao score for predicting college success. Furthermore, Huang's team investigated the equity implications of the test between rural and urban residents, males and females, and between "super high school" and the regular school students. In addition, the team examined the predictive validity and social consequences of University K's independent recruitment examination. Their results suggested that (a) Gao-Kao score was a fairly good predictor of the freshmen grade; (b) the score usage had a negative impact on rural students; (c) the performance by different gender group was similar; (d) "super high schools" students were very much advantaged; however, students from these high schools did not perform significantly better in their first college year; (e) University K's independent recruitment efficiently selected students with better academic performances and achieved better gender balance, but was even more favorable to urban residents than Gao-Kao. In the new round of "Gao-Kao" reform, Prof. Huang suggests a monitoring system on the impact of test score usage be established, through which policy makers may find the best balancing point between efficiency and equity.

In PROMS 2018 held on campus of Fudan University, Professor Jin Yan from Shanghai Jiaotong University was invited to present a keynote speech addressing "Testing Tertiary-Level English Language Learners: The College English Test in China".

The CET is an exit test to assess the English language proficiency of tertiary-level learners upon their completion of the English language education in colleges and universities in China. In this presentation, Jin presented an account of the development of the CET in the past three decades. The purpose is to demonstrate the value as well as the challenges of developing and using a contextualized national test to promote English language education at the tertiary level in China. In this keynote presentation, an overview of the CET is provided through reflection on how and why this testing program started in the 1980s and the subsequent milestones in the evolution of the test. The localized nature of the CET testing system is then discussed in relation to those features unique to the educational and assessment context in China. The measures taken by the CET test developers and administrators to control the quality of the test and its fairness are again reviewed. Innovations in technology-enhanced assessment practices are introduced, including the computer-based CET Spoken English Test and the internet-based CET, automated scoring of CET essay writing and paragraph translation, and counter-cheating techniques. Following the review of test development, an evaluation of the uses of the CET and its impact on teaching and learning are carried out. Next, the efforts made by the test developers and independent researchers on conducting evidence-based validation of the CET are addressed in details. Finally, the speaker discussed the future directions in research and practice on the largest-scale English language test in the world.

Followed by Jin's speech, Prof. Bond delivered his keynote titled "What Does Rasch Measurement Offer English Language Testing in the Orient?" This time, Bond's speech was inspiring because it is the right person to declare on the occasion that the "Rasch Wars" (see Section 1.6) in language testing are both over and won. In Bond's opinion, the Rasch Wars have had little direct impact on language testing practices in the orient (He mainly refers to China). To him, it seems that the orient remains a key market for "international" language tests from the USA and UK. Local language practitioners report large gaps between the tests delivered by those companies and what they see as the needs of local institutions, students and teachers. This keynote addresses three main points: (a) describes how small groups of language teachers developed their own tests of English as a second language; (b) reports on the psychometric properties of an English language test at a prestigious Chinese university; and (c) suggests strategies for improving high-stakes language testing in China. Needless to say, the common thread running through these projects is the use of methods and principles derived from Rasch measurement theory.

In PROMS2023, Macau, China, Professor Yaru Meng[15] from Xi'an Jiaotong University, China was invited to deliver a keynote entitled: Concentrated on "Bug-CDM Based Diagnosis of EFL Learners' Listening Barriers through MCQ Incorrect Options". Prof. Meng's keynote addresses Cognitive Diagnostic Models (CDMs) and multidimensional latent variable models for cognitive diagnosis. Meng also demonstrated their advantages in English as Foreign Language (EFL) listening in identifying learners' cognitive processes in test items, and offering feed-back of their strengths and weaknesses. According to Meng, this is achieved by adopting appropriate CDMs to analyze Q-matrix (of test item and the examined attributes) and learners' dichotomous responses of 0 or 1. The results revealed that Bug-GDINA may be the optimal model. The study proved the feasibility of Bug-CDMs (Bug-GDINA in particular) in diagnosing listening problems from the incorrect options. Limitations and potentials for future research are also proposed.

In the similar fashion, Dr. Zouh Fong Chieng[16] presented the topic: Rasch's Distractor Analysis for Multiple-Choice Questions at PROMS 2024, Kuala Lumpur, Malaysia. This study aims at applying Rasch model distractor analysis for dichotomous data in educational assessment. Reliability and validity are the prime concern aspects for multiple-choice questions (MCQ) item design. Beyond it, the efficiency of distractors become the next most important challenge. Quality of MCQ items is traced back to the qualities of available distractors. This plays a crucial role in formulating the correct responses from examinees and ensuring appropriate levels of item difficulty. Zouth's study focuses on addressing distractors that cause model misfit, with the goal of enhancing the measurement qualities of an instrument by reducing measurement error and item misfit. The main objective is to ensure that each distractor functions correctly, effectively discriminates between students who have mastered the content and those who have not. Zou's research is a quantitative research design with empirical data from Malaysian School Assessment. An analysis of distractors in principle of accounting subject is described. The instrument consists of 40 MCQ items with four options, that is, one correct response and three distractors. The sample size consists of 429 students chosen as respondent for Text Form-X (XQ) and of another 429 students for Test Form-Y (YQ). The data were initially scored dichotomously, and missing responses were scored as incorrect. Analysis of distractors was plotted in graphical display. To understand more of mis-functioning distractors which cause the model misfit, the content was examined by the subject matter expert. Analysis according to the model will reveal whether the intended order of the distractors functions empirically as expected. An examination of distractors holds particular significance in teaching and learning as it unveils misunderstandings and misconceptions, notably among lower proficiency subject groups. Zouh concluded that future studies would concentrate on other methods in detecting mis-functioning distractors. This would help educators improve the quality of assessment items and gain insights into students' understanding of the material being assessed.

Dr. Bambang Sumintono,[17] Faculty of Education, Universitas Islam Internasional Indonesia, was invited to deliver the keynote entitled "Are First Presenters Assessed Fairly? Application of Many Facet Rasch Model in Proposal and Thesis Examination" Bambang addressed that students' proposal and thesis examinations in a university are the formal assessment wherein they show an independent research project and are reviewed by a panel of examiners. The process of evaluation usually involves using assessment rubric form that fills it by examiners called as multi-rater. In his speech, Bambang analyzed the multi-rater data of students' proposals and thesis examinations of a master degree program using Many Facet Rasch Model (MFRM). The sample size Bambang used consisted of twenty students being assessed by eleven lecturers as raters. There were two components for each stage which are presentation and written examination. In this research, Bambang obtained two interesting findings: (a) for both proposals and thesis examinations, students always got better scores from oral presentation than from written form. (b) In thesis examinations, students who are voluntarily assessed earlier tend to get lower marks as proof by bias interaction report. This study shows the importance of moderation between raters in order to get fair assessment. Dr. Haniza[18] Yon from Malaysian Psychometrics Association (MPA) was invited to make the keynote speech on "Improving Fairness in Educational and Psychological Testing Using Computerized Adaptive Testing (CAT)". Haniza addressed that psychological assessment, aptitude and knowledge testing of individuals has a well-established history over many centuries. Owing to the strong influence of the US civil rights movement, the second half of the last century saw a strong push toward establishment of well-defined criteria for test equating and overall fairness—*even when over time different items were used to assess the same content.* Although these criteria are not accepted worldwide, they have become standard in most developed countries. While improvement and extension of testing theory and applications occurred rather slowly in the past, today we see revolutionary developments at a rapid pace, due to two major technical advances. The first of these concerns **basic computer technology**. The advent of affordable online computer systems and the possibility of creating (nearly) equivalent tests with different items gave rise to CAT, a powerful method in which a computer program administers test questions according to a dynamic algorithm that continuously estimates the ability level of the test taker and chooses successive questions accordingly. The second technological advance involves **large language modeling (LLM)**, which interfaces with many types of psychological and educational testing and enables augmentation of standard CAT. In this speech, Haniza discussed both areas of technical advancement and how CAT can lead to improved sophistication in educational and psychological testing.

6.3.3 Creative Thinking

In PROMS 2013 organized by National SunYat-sen University, Kaohsiung, China, Dr. Jing-Jyi Wu[19] from National Chengchi University, China was invited to deliver the speech entitled "Assessing Creativity, Then and Now: Some Examples from Chinese Society". Professor Wu's speech should be considered as one of the earlier studies on evaluating creativity. Wu addressed that throughout history, creativity and innovation have been very important, especially in education. Nowadays in the increasingly complex, dynamic, and globalized world, they are more important than ever. He drew an impressive example. In 2010, IBM's annual CEO survey, collecting responses from more than 1500 CEOs in 33 fields across 60 countries, showed that the number one most important skill required for leadership while facing the dynamic challenges of the world today is creativity. Wu emphasized that creativity is more important than ever, as is the need to assess creativity, such as for the selection of creative talent or evaluation of creative products. The emphasis on the need to assess creativity is contrasted with standardized testing, which is a barrier to creativity. He went on, in Chinese societies, entrance examinations, direct teaching and student evaluations, and ultimately stifle creativity. Historically, this traces back to the Ke-Ju (i.e., imperial) examination system in China that, while providing more opportunity for advancement, also fostered uniformity and memorization, such as through the uncreative Ba-Gu-Wen (i.e., the eight-legged essay[20]). In terms of Big-C creativity in STEM areas, the Chinese often cite the Four Great Inventions (gun powder, compass, papermaking, and printing, but as the famous Needham[21]

Question shows, creativity has not remained a defining characteristic of Chinese society. The formal examination system for selecting officers is criticized as creating a culture deficient in creativity.

According to Wu, outside formal assessment, however, communities of scholars in the arts, especially literature, practiced informal peer evaluation that helped identify and develop creative talent and works. These long-standing consensus-making methods were the predecessors of the consensual assessment technique creativity scholars are familiar with today. A third historical area of assessment was that of everyday or little-c creativity, as told through anecdotes of how creative people solve everyday problems in informal and formal contexts.

In Wu's viewpoint, assessing creativity in formal education gained traction after many gatekeepers in education and the government returned from studies in the USA. The need to develop approaches for assessing creativity was again voiced alongside criticisms of creativity-killing entrance examinations. It was within this climate that creativity tests were translated or newly developed.

Wu concluded that society in general still needed mechanisms for the selection of creativity, especially in non-mainstream areas, like theatre or the arts, in informal contexts. Parallel to history, outside of the formal education system, creativity was identified without formal creativity tests but again through alternative forms of consensual assessment of authentic performance, such as auditions for theatre selections and evaluation of creative performances. Anecdotes of creativity from children to adults, and among artists, scholars, and scientists were introduced either in speech or publications, giving the general public a chance to understand how creativity is assessed informally in everyday life.

In PROMS2023, Macau, China, Professor Jue Wang[22] from University of Science and Technology of China, Anhui, China was invited to deliver a keynote entitled: Measuring Scientific Creativity with Subjective Creativity Assessments: Psychometric Challenges and Suggestions. In her speech, Wang addressed that the research on scientific creativity plays an increasingly important role in education. Scientific creativity can be viewed as a type of domain-specific creativity. Subjective creativity assessments (SCA) are particularly useful for measuring domain-specific creativity and earned wide popularity in evaluating product-based creativity. However, under the psychometric framework of examining rating scores, the application of SCA has several challenges, including (a) a lack of clarity and consistency among raters on the understanding and rating criteria of scientific creativity, (b) varying levels of reliability and validity across studies, and (c) insufficient evidence for supporting fairness and comparability arguments. In this study, Wang addressed several issues related to raters and rating scores with the use of SCA and provided a psychometric framework for evaluating the quality of ratings and assessment design of SCA. Suggestions for psychometric analysis for creating objective and fair measurement of scientific creativity are discussed as well.

6.3.4 Technical Issues of the Application of Rasch Model

In PROMS 2018 organized by Shanghai Fudan University, Dr. Xun Yan[23] from University of Illinois at Urbana-Champaign, USA was invited to make a featured speech regarding "How Does Rater Performance Change over time? Insights Gained from Many-Facet Rasch Modeling in Second Language Writing Assessment". Xu addressed that in language testing, rater studies tend to examine rating performance at a single time point. Research on the effect of rater training either adopts a pre- and post-training design or tracks rater performance across multiple time points within a short period of time. Xu examined how rater performance changed during a semester-long rater training and certification program for a post-admission English as a second language writing placement test at a large US university. According to Xu, the certification program aims to align raters to a newly developed rating scale that provides both placement recommendations and diagnostic information regarding students' writing proficiency level and skill profile. The training process employed an iterative, three-stage approach, consisting of face-to-face group meetings, individual rating exercises, and scale re-calibration based on rater performance and feedback. Using many-facet Rasch modeling (Linacre, 1989, 2006), Xu analyzed rating quality of 17 novice raters across four rounds of rating exercises over the course of a semester. Rating quality was operationalized in terms of rater severity and consistency, rater consensus, and raters' use of rating scale. These measurement estimates of rater reliability were compared across time and between certified and uncertified raters. Xu found out that at the start of the training program, all raters were inconsistent, varied widely in severity, and achieved low exact score agreement. Over time, certified raters improved on multiple measures of reliability and became more indistinguishable with each other in the application of the rating scale. However, rater performance did not improve in a linear fashion but instead followed a U-shaped developmental pattern. In contrast, uncertified raters' performance remained similar across rounds. The findings obtained by Xu have implications for the effectiveness of rater training and developmental patterns of rating behavior over time.

In PROMS2021, Nanjing, China, Prof. Steven from Wesleyan University, USA was invited to deliver the keynote speech addressing "Better Measurement and Fewer Parameters! The True Value of Rasch over IRT!" Steven's ideas were applauded by the online attending audience. Under the background of the rapid development of network technology in the era of AI, the present author believes that Rasch model is a right model. The Rasch-based research direction will focus on more correct rather than on more accurate!

In PROMS2023, Macau, China, Professor Guangzhong[24] Luo from Jiangxi Normal University, Nanchang, China was invited to deliver a keynote entitled: The JML estimation in the Equi-distant Unfolding model for Polytomous Responses. This is a speech that requires a high level of professional expertise. Luo operationalizes the general unfolding model for polytomous responses under the condition of equal distances between the successive thresholds of an item. Using the JML estimation, two algorithms are proposed for parameter estimation, each with a different technique to reduce the bias in the estimates produced by the inconsistency of the JML. One algorithm is a direct extension of the algorithm (Luo et al., 1998) in which a correction is applied on the mean latitude of acceptance parameter. Simulation studies show the efficiency of the algorithms in various situations. An analysis of real data thus collected from questionnaire based on 380 students from Australia, Japan, and Singapore is included for illustrative purposes (Andrich and Luo, 2003).

Luo argued that though Thurstone's (1928) scaling principles were widely accepted as rigorous, the most popular method for the measurement of attitude in the last 70 years was not Thurstone's, but Likert's (1932) summated rating scaling. In fact, after reversing the responses on the negative items, Likert's scaling uses a cumulative process characterized by monotonic response functions. The total scores of persons across items are used as the measures of attitudes. In contrast, the mid-point principle, which is used in the Thursone's scaling, implies a single-peaked response function which characterizes what is often termed an unfolding process. According to Luo, these two response processes are not compatible (Andrich, 1996; Roberts et al., 1999). The algorithm[25] Luo addressed in the speech provides another example of the operationalization of unfolding models for statements with polytomous response formats. Though more investigation on parameter estimation would help (a comparison study on the estimation result of JML and MML, for example), the ideas presented on this occasion show that this operationalization efficiently recovers the parameter values of the model with satisfactory accuracy.

6.3.5 Next Generation Measurement: When Innovation Meets Objectivity

The theme of the 20th PROMS (2025) Singapore is "Next Generation Measurement: When Innovation Meets Objectivity". The theme emphasizes the commitment to objectivity or invariance in the field in response to cutting-edge advancements and emerging technologies that reshape how the Raschers collect, analyze, and interpret data. This theme invites a critical examination of the ways innovation can enhance precision, accuracy, and fairness in measurement practices across diverse disciplines. The 20th PROMS (2025) will serve as a platform for experts, researchers, and practitioners to share insights and best practices, fostering dialogue on the future of measurement that is both forward thinking and ethically grounded. Submissions include research applying the Rasch model, advances in measurement practices, or reviews of modern measurement theory. Topics of submission may include but are not limited to the following:

Integrating Rasch measurement with AI

- Using Rasch for personalized learning and feedback
- Advances in Rasch modelling

- Advances in the practice of measurement
- Issues relating to modern measurement theory
- Technical issues of the application of Rasch model
- Measurement for advancing scientific fields
- Measurement across cultures
- Measurement of soft skills and competencies
- Measurement technology
- Questionnaire and test development
- Rater-mediated assessment
- Fairness of using assessment to make decisions
- Assessing language skills

Apart from this, PROMS2025 has an illustrious group of speakers, below are a few highlights:

Emeritus Professor David Andrich, The University of Western Australia. This distinguished scholar presentation will address: Rasch Metameters of Growth in Reading and Mathematics: Attainment at Both Population and Individual Levels;

Professor George Engelhard Jr., University of Georgia, USA. George will present his presidential address: Measurement, Explanation, and Invariance: Next Generation Invariant Measurement;

Professor Jue Wang, University of Science and Technology of China. Wang's keynote presentation will be addressing Human Scoring in Subjective Creativity Assessment;

Associate Professor Vahid Aryadoust, Nanyang Technological University, Singapore. Vahid's keynote presentation will focus on: What Can Generative AI Do for Assessing Listening and Speaking Skills';

Dr Lye Che Yee, Singapore University of Social Sciences. Lye's keynote speech will present: Refocusing Educational Measurement: Understanding Student Learning through Adaptive Learning.

These keynote speeches are novel and refreshing to the international attendees. The convening of PROMS conferences has indeed provided researchers, scholars, practitioners, MA, and PhD students with an excellent opportunity for academic exchange and learning, and in turn has given foreign peers a comprehensive understanding of the time and the in-depth research, application, and development of the Rasch model in the Pacific-Rim countries and regions and beyond, which has played a greatly positive role in promoting further learning in the future. Table 6.3 provides the detailed list of PROMS Conference and keynote speeches titles over the past two decades.

TABLE 6.3

List of PROMS Conferences and Keynote Speeches Over the Past 20 Years

Years and Venues	Speech Titles	Keynote Speakers	Universities
The 1st PROMS 2005, Malaysia Dates: June, 2005 Venue: International Islamic University of Malaysia (IIUM) Kuala Lumpur, Malaysia Hosted by Prof Noorlide Abu Kasim and Prof. Nor Akbariah Makhzir	Achieving the impossible: Solving real measurement problems with the Rasch model	Prof. Trevor G. Bond	James Cook University, Australia
The 2nd PROMS 2006, China Venue: The Hong Kong Institute of Education, Hong Kong, SAR, China	The development and application of Rasch measurement in educational contexts	Prof. Mark Wilson	University of California, Berkeley (UCB)
	The practical applications of Rasch measurement models in the field of educational assessments, with insights from the Hong Kong Examinations and Assessment Authority.	Dr. Ida Marais	Hong Kong Examinations and Assessment Authority, Hong Kong, China
The 3rd PROMS 2007 Dates: 5–7 August 2007 Venue: National Chung Cheng University, Chiayi, Taiwan, China	Rasch Modelling as a Framework for Integrating Assessment and Instruction	Prof. Trevor Bond	James Cook University, Australia
	Invariant Measurement: Using Rasch Models in the Social, Behavioral, and Health Sciences.	Prof. George Engelhard, Jr.	University of Georgia, USA
The 4th PROMS 2008 Dates: 31 July–3 August 2008 Venue: Ochanomizu University Tokyo, Japan **Conference Theme:** *Objective Measurement for Teaching, Learning*	Rasch Measurement: The Path to Objectivity	Prof. Benjamin D. Wright.	University of Chicago, USA
	Cognitive Assessment: Cognitive Diagnosis and Q-Matrix Theory	Prof. Kikumi K. Tatsuoka	Columbia University New York, USA

(*Continued*)

TABLE 6.3 (*Continued*)

List of PROMS Conferences and Keynote Speeches Over the Past 20 Years

Years and Venues	Speech Titles	Keynote Speakers	Universities
The 5th PROMS 2009 Dates: 28–30 July 2009 Venue: The Hong Kong Institute of Education, Hong Kong, SAR, China Conference Theme: Quality Rasch Measurement	Constructing Measures: An Item Response Modeling Approach	Dr. Mark Wilson	University of California, Berkeley (UCB)
	Rasch Models and the Measurement of Change	Dr. Trevor Bond	James Cook University, Australia
	The Role of the Rasch Model in the Development of a Learning Progression	Dr. Stephen Humphry	The University of Western Australia
The 6th PROMS 2010 Dates: 29 June–1 July 2010 Venue: International Islamic University, Kuala Lumpur, Malaysia	The Challenge of Measuring Abilities and Attitudes	Dr. Trevor Bond	James Cook University, Australia
	The Many-Facet Rasch Model: Theory and Application	Dr. Ida Marais	Hong Kong Examinations and Assessment Authority, Hong Kong, China
	A Course on Rasch Measurement Theory and Its Applications	Dr. David Andrich	The University of Western Australia
The 7th PROMS 2011, Singapore Dates: 13–15 July 2011 Venue: National Institute of Education, Singapore *Conference Theme: Objective Measurement for Teaching, Learning and Living*	The Rasch Model as a Statistical Model	Prof. David Andrich	The University of Western Australia
	The Challenges of Applying the Rasch Model for Measurement in Health Outcomes	Prof. Trevor Bond	James Cook University, Australia
	The Role of Measurement in Education	Prof. Geoff Masters	Chief Executive Officer Australian Council for Educational Research (ACER)
The 8th PROMS 2012, Jiaxing, China Dates: 6–9 August Venue: Jiaxing University, China The first time in China Mainland	Five Steps to Rasch: Simplifying the Construction of Valid Measures	Prof. Trevor G. Bond	James Cook University, Australia
	Rasch Modelling in the Analysis of Self-Reported Survey Data: Uses, Benefits, and Challenges	Prof. Rob Cavanagh	School of Education, Curtin University, Australia

TABLE 6.3 (*Continued*)

List of PROMS Conferences and Keynote Speeches Over the Past 20 Years

Years and Venues	Speech Titles	Keynote Speakers	Universities
	Rasch Analysis for Psychological and Educational Measurement: Recent Advances and Future Directions	Prof. Wen Chung Wang	The Hong Kong Institute of Education, Hong Kong, SAR, China
	On the Potential for Improved Measurement in the Human Social Sciences	Prof. Jack Stenner	Chairman and CEO and Co-founder Meta Metrics, USA
	The Ten-Year Matriculation English Test (MET) Equating Project via the Rasch model	Professor Quan Zhang	Jiaxing University, China
	Rasch, Complexity, and the Science of Measurement	Dr. William F. Fisher	Graduate School of Education, University of California, Berkeley, CA, USA
The 9th PROMS 2013, Kaohsiung, China Dates: 3–5 August 2013 Venue: National Sun Yat-sen University, Kaohsiung, China	Measurement must balance its statistical and scientific elements	Prof. Mark Ronald Wilson	University of California, Berkeley, USA
	Assessing Creativity, Then and Now: Some Examples from Chinese Society	Dr. Jing-Jyi Wu	National Chengchi University, China
	Combining Adaptive Testing with Adaptive Learning	Dr. Hua-Hua Chang	University of Illinois at Urbana-Champaign, USA
	The Assessment and Teacing of 21st Century Skills	Dr. Patrick Griffin	University of Melbourne, Australia
	Language Testing in China: The Space between Theory and Practice	Prof. Quan Zhang	Jiaxing University, China
	Evidence and Quality Assurance in the Finnish Educational System	Prof. Hannele Niemi	University of Helsinki, Finland

(*Continued*)

TABLE 6.3 ***(Continued)***

List of PROMS Conferences and Keynote Speeches Over the Past 20 Years

Years and Venues	Speech Titles	Keynote Speakers	Universities
	The Development and Applications of DIF-free-then-DIF Strategy	Dr. Ching-Lin Shih	National Sun Ya-sen University, Taiwan, China
	Performance=based versus Objective Content-based Item Analysis: Quantitative Syntactic Network Perspectives	Dr. Chi-Chien Yang	National Taichung University of Education, Taiwan, China
The 10th PROMS 2014, Guangzhou, China Dates: 4–6 August 2014 Venue: Clifford Hotel, Panyu, Guangzhou, China Hosted jointly by Jiaxing University and City Training Institute (CTI), Guangzhou, China	An Argument Approach to Test Fairness: The case of multiple-form equating in the College English Test	Prof. Yan Jin and Dr. Eric Wu	Shanghai Jiaotong University, China University of California, Los Angeles (UCLA), USA
	A System for Diagnosing and Tracking English Competence in Chinese University Students	Alan Urmston Michelle Raquel Gwendoline Guan	Hong Kong Poly U, Hong Kong, China
The 11th PROMS 2015, Fukuoka, Japan Dates: 20–24 August 2015 Venue: Kyushu Sangyo University Fukuoka, Japan	The Past, Present, and Future of Rasch Measurement	Prof. Trevor Bond	James Cook University, Australia
	Applying the Rasch Model to the Assessment of Health Outcomes	Prof. Everett Smith	University of Illinois at Chicago, USA
	Rasch Models in the Field of Education: Challenges and Opportunities	Prof. Ryosuke Ohara	The University of the Ryukyus, Okinawa, Japan
The 12th PROMS 2016 Xi'an, China Dates: 30–31 July 2016 Venue: Oriental Hotel, Xi'an, China Hosted jointly by Jiaxing University and City Training Institute (CTI), Guangzhou, China	Exporting to China: The future of a Genuine Collaboration with the West	Prof. Trevor G. Bond and Prof. Zi Yan	James Cook University, Australia The Education University of Hong Kong, SAR, China
	Investigating the Predictive Validity and Social Consequences of "Gaokao"	Professor Xiaoting Huang	Peking University, China

TABLE 6.3 (*Continued*)

List of PROMS Conferences and Keynote Speeches Over the Past 20 Years

Years and Venues	Speech Titles	Keynote Speakers	Universities
	Substantive Theory General Objectivity and an Individual-Centered Psychometrics	Prof. Jack Stenner	Chairman and CEO and Co-founder Meta Metrics Inc., USA President of the Board of Directors of Institute of Objective Measurement, a board member for the National Institute for Statistical Sciences, and a past board member for Duke Children's Hospital and the North Carolina Electronics and Information Technologies Association, USA
	Applying the Rasch Model in an Operational Language Testing Program	Dr. Jamie Dunlea	Language Assessment Research Group at the British Council, London
	Development of an Automated Scoring System for English Essays Using Latent Semantic Analysis	Dr. Haniza Yon	President of the Malaysian Psychometrics Association
	Multifaceted Rasch Analysis in Rating Tasks in Japan's University Entrance Examination	Prof. Rie Koizumi	Juntendo University, Japan Secretary General of the Japan Language Testing Association (JLTA)
The 13th PROMS 2017 Kota Kinabalu, Malaysia Dates: 5–9 August 2017 Venue: Universiti Malaysia Sabah Malaysia	Rater Effects and IRT models	Prof. Margaret Wu	University of Melbourne, Australia
	Invariant Measurement and Metrological Networks in a Modern Measurement	Prof. Robert Frederick Cavanagh	Curtin University Western Australia
	Rasch in Malaysia—A Brief History, Challenges, and a Peek into Future	Dr. Mohd Zali Mohd Nor	myRasch Malaysia

(*Continued*)

TABLE 6.3 (*Continued*)

List of PROMS Conferences and Keynote Speeches Over the Past 20 Years

Years and Venues	Speech Titles	Keynote Speakers	Universities
The 14th PROMS 2018, Shanghai, China Dates: 23–27 July 2018 Venue: Fudan University, China Hosted by Fudan University, China	Fairness and Justice in Language Assessment: What Role Can Rasch Measurement Play?	Prof. Tim McNamara	The University of Melbourne, Australia
	What Does Rasch Measurement Offer English Language Testing in the Orient?	Prof. Trevor G. Bond	James Cook University, Australia
	Testing Tertiary-Level English Language Learners: The College English Test in China	Professor Jin Yan	Shanghai Jiaotong University, China
The 15th PROMS 2019, Surabaya, Indonesia Dates: 26–31 August 2019 Venue: Bumi Surabaya City Resort Hosted by Indonesian Association of Psychometrics (Apsimetri) & Faculty of Psychology, Universitas Airlangga, Surabaya, Indonesia	Applied Psychological and Rasch Model in Psychological Measurement and Educational Testing	Rob Cavanagh, PhD	Curtin University, Australia
		Prof. Jeffrey Durand	Toyo Gakuen University, Japan
		Prof. Bambang Sumintono	Universitas Malaya, Kuala Lumpur, Malaysia
		Jahja Umar, PhD	UNI Sportif Hidayotullah, Indonesia
		Urip Purwono., M.S., M.Sc., Ph.D.	Universitas Padjadjaran, Bandung, Indonesia
		Bahrul Hayat PhD	UNI Sportif Hidayotullah, Indonesia
The 16th PROMS 2021, Nanjing, China Dates: 4–6 December 2021 Venue: Nanjing Normal University, China The first large-scale online PROMS during the global fighting against COVID-19 Pandemic	From Estimation to Consideration: The Role of Rasch Measurement in Promoting Understanding and Validity	Prof. Trevor G. Bond	James Cook University, Australia

TABLE 6.3 (*Continued*)

List of PROMS Conferences and Keynote Speeches Over the Past 20 Years

Years and Venues	Speech Titles	Keynote Speakers	Universities
	Large-scale International Educational Assessment: Uses, Limitations, and Counter-Intuitive Findings	Prof. Kit Tai HAU	The Chinese University of Hong Kong
	Response Styles as Person Differential Functioning: Methodological Approaches to Solve Person DIF	Prof. Ricardo Primi	São Francisco, Brazil
	Better Measurement and Fewer Parameters! The True Value of Rasch over IRT	Prof. Steven Stemler	Wesleyan University, USA
The 17th PROMS 2022, Hanoi, Vietnam			
The 18th PROMS2023, Macau, China Dates: 29–30 August 2023 Venue: The World Sports University Macau, SAR, China	Invariant Measurement and the Three Traditions of Measurement in the Human Sciences	Prof. George Engelhard Jr.	The University of George, USA
	How Rasch Models Work as AI Engine in Assessment	Prof. Luo Guanzhong	Jiangxi Normal University, China South China Normal University, China
	Investigating University Student's Collaborative Problem Solving Competency Taking the Computational Psychometrics Approach	Prof. Huang Xiaoting	Peking University, China
	Rater Issues in Subjective Creativity Assessment: Psychometric Challenges and Future Directions	Prof. Jue Wang	University of Science and Technology of China, China
	Concentrated on "Bug-CDM Based Diagnosis of EFL Learners" Listening Barriers through MCQ Incorrect Options	Professor Yaru Meng	Xi'an Jiaotong University, China

(*Continued*)

TABLE 6.3 (*Continued*)

List of PROMS Conferences and Keynote Speeches Over the Past 20 Years

Years and Venues	Speech Titles	Keynote Speakers	Universities
The 19th PROMS2024, Kuala Lumpur, Malaysia Dates: 19–21 August 2024 Venue: University College MAIWP International	Mapping the Rating Mind: Comparison of Rating Categories in Personality and Motivation Measures	Prof. Harris Shah Abd Hamid	University College of MAIWP International & Universitas Negeri Malang, Malaysia
	Reliability Is Straightforward; Validity Is Not So Easy: Using the Rasch Model for Test Development, Maintenance and Implementation.	Prof. Trevor G Bond	James Cook University, Australia
	Are First Presenters Assessed Fairly? Application of Many Facet Rasch Model in Proposal and Thesis Examination	Prof. Bambang Sumintono	Faculty of Education, Universitas Islam Internasional Indonesia
	Improving Fairness in Educational and Psychological Testing Using Computerized Adaptive Testing (CAT).	Dr. Haniza Yon	Malaysian Psychometrics Association
The 20th PROMS 2025, Singapore Dates: Venue: Singapore University of Social Science (SUSS), Singapore	Rasch Meta-meters of Growth in Reading and Mathematics: Attainment at Both Population and Individual Levels	Emeritus Professor David Andrich,	The University of Western Australia
	Presidential Address: Measurement, Explanation, and Invariance: Next Generation Invariant Measurement	Professor George Engelhard Jr.	University of Georgia, USA
	Addressing Human Scoring in Subjective Creativity Assessment	Professor Jue Wang,	University of Science and Technology of China
	What Can Generative AI Do for Assessing Listening and Speaking Skills	Prof. Vahid Aryadoust,	Nanyang Technological University, Singapore
	Refocusing Educational Measurement: Understanding Student Learning through Adaptive Learning	Dr Lye Che Yee	Singapore University of Social Sciences

6.4 The Development and Impact from PROMS

Ever since the first PROMS was held in Kuala Lumpur, Malaysia in 2005 as well as the one in Jiaxing, China Mainland in 2012 all with great successes, the convening of the conferences has greatly contributed to popularizing, promoting, exchanging, and improving the research and application of the Rasch measurement theory and application. This is mainly reflected in the following five aspects.

6.4.1 Participants and the Growth of the Rasch-based Publication

As shown in Figure 6.1, two significant points of time (2005 and 2012) are, respectively, the commencement of PROMS 2005, Kuala Lumpur, Malaysia indicating the increasing growth of the publication of the Rasch measurement research and the beginning of PROMS2012 Jiaxing, China reveals the remarkable growth of the publication of the Rasch measurement research (Bond and Yanzi, 2016).

And the participants attending PROMS conferences come from Australia, Brazil, China Hong Kong, China Macau, China Taiwan and China Mainland, Denmark, Indonesia, Japan, Malaysia, Philippines, Singapore, South Korea, Sweden, Vietnam, USA, UK, and far beyond Pacific-rim regions and countries. After the successful PROMS2022 Hanoi, Vietnam followed by PROMS2023, Macau, China, and PROMS 2014, Kuala Lumpur, Malaysia, Singapore, will open their arms to embrace PROMS2025.

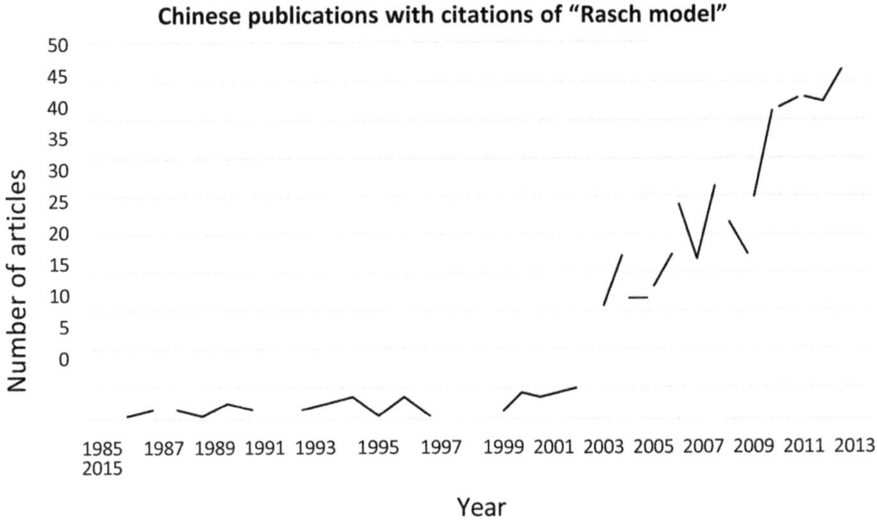

FIGURE 6.1
Citations of applying the "Rasch model"

6.4.2 Conference Proceedings

Today, not all large international academic conferences publish conference proceedings, and even if there are, they are rarely published continuously. The PROMS conference proceedings are officially published by the internationally renowned publisher the Springer and included in databases such as CPCI, CNKI, Google Scholar, etc. for six consecutive years (Zhang, 2013, 2015, 2016, 2018, 2024, the forthcoming 2026). The paper-based PROMS proceedings have been collected in major libraries and universities around the world. The e-books are included in the Springer Link Behavioral Science and Psychology e-book collection. This is accessible to readers around the world. The downloading and citations are increasing annually.

6.4.3 Rasch Monographs

Led by Prof. Magdalena Mo Ching MOK from Hong Kong Educational University and Prof. Quan Zhang from Jiaxing University and with joint efforts, the series translation of books has been published. They are: Constructing variables. Book of Abstracts Vol. I and II, *Journal of Applied Measurement*, JAM Press USA, 2014 and 2015. Both books contain 205 abstracts germane to Rasch-based research work. This qualified team of translators consists of 45 Rasch experts with highly bilingual proficiency working in 13 different organizations located in different parts of China.

Each of these abstracts deals with Rasch measures in their research field, covering a variety of topics. They serve as good resources for Chinese-speaking researchers and students of non-English majors to be able to conduct their own Rasch-related research (Zhang and Zhang, 2014).

David Andrich and Ida Marais (2019) published their monograph entitled: A Course in Rasch Measurement Theory: Measuring in the Educational, Social and Health Sciences, the Springer Nature. The book applies Rasch measurement theory to the fields of education, psychology, sociology, marketing, and health outcomes in order to measure various social constructs. The chief focus is on first principles of both the theory and its applications. As the software they used is RUMM (see Section 3.8.6), which is readily available to carry out analyses of real data, numerous small examples are provided in the book. The goals are to equip researchers with the confidence in order to be in control of the analysis and interpretation of data, and to make professional rather than primarily statistical decisions mechanically. The content is based on courses that have been taught both online and in intensive form for over two decades. Much of the book is based on research conducted by the two authors and their colleagues.

So far totally four editions of the monograph "Applying the Rasch Model: Fundamental Measurement in the Human Sciences" by Trevor G. Bond and his co-authors (2021) have been published by Routledge, Taylor & Francis Group. Recognized as the most influential publication in the field, the books facilitate deep understanding of the Rasch model and provide practical applications. Bond and his disciples review the crucial properties of the model and demonstrate its use with examples across the human sciences. In particular, the glossary and illustrations support that understanding, and the accessible approach means that it is ideal for readers without a mathematical background. This is what readers are most enthusiastic about. Highlights of the newest edition include major contributions such as more learning tools to strengthen readers' understanding, greater emphasis on the use of R packages, good distinction between numerical values, quantity and units to understand the measurement and the role of the Rasch logit scale, and the clarified relationship between Rasch measurement, path analysis and SEM, with new examples of Rasch measurement applied across health sciences, education and psychology. George Engelhard (2013) published Invariant Measurement Using Rasch Models in the Social, Behavioral, and Health Sciences by Routledge, Taylor & Francis Group. And eleven years later, the 2nd Edition (2024) was published jointly by George Engelhard and his PhD student, Prof. Jue Wang from University of Science and Technology of China, China. In George's book, three key points are fully addressed, i.e., (1) the principles of invariant measurement, (2) how invariant measurement can be achieved with Rasch models, and (3) how to use invariant measurement to solve measurement problems in the social, behavioral, and health sciences. Throughout the book, Rasch models are used but a comparison of Rasch models to IRT models was also described. Written with students in mind, the book was class tested to help maximize accessibility. Six highlights include: a strong philosophical and methodological approach to measurement in the human sciences; good demonstrations of measurement problems using invariant measurement; practical illustrations using invariant measurement; a history of measurement based on test-score and scaling traditions; the detection and measurement of rater errors and the evaluation of rater accuracy, and a review of estimation methods, model-data fit, indices used to evaluate the quality of rater-mediated assessments, rater error and bias, as well as rater accuracy.

More recently, A.Y.M. Atiquil Islam (2023) published Applying the Rasch Model and Structural Equation Modeling to Higher Education by CRC Press, Taylor and Francis. Atiquil's book introduces the fundamentals of the technology satisfaction model (TSM), supporting readers in applying the Rasch model and SEM, a multivariate technique to higher education (HE) research. The TSM addressed in the book includes digital technologies for teaching, learning and research across three dimensions: (1) computer efficacy, (2) perceived ease of use, and (3) perceived usefulness. According to the author, establishing such relationships among these factors is a challenge and the author successfully shows that combining these two analytical tools offers Rasch counterparts better options for measurement and generalization in HE research.

6.4.4 Participants

Academically, the PROMS2014 symposium tends to be more international. There are at least four highlights. The first one is physical. The number of participants is increasing, and the academic impact becomes obvious in the

TABLE 6.4

Sponsors, Donors and Supports for PROMS Conferences over the Past Two Decades

Working Units	Years	Countries/Regions
Atlantic Press, part of /Springer Nature	2024	German
Beijing Higher Education Press	2012, 2014	China
The British Embassy Cultural and Educational Association	2012	UK
City Training Institute (CTI) Guangzhou	2014, 2016, 2023	China
Correction Network	2018	China
The Educational Testing Service (ETS)	2012	USA
Foreign Language Teaching and Research Press	2012	China
Fudan University	2018	China
Hong Kong Institute of Education	2006, 2009	China Hong Kong
International Islamic University of Malaysia (IIUM)	2005, 2010	Malaysia
Indonesian Association of Psychometrics (Apsimtri & Faculty of Psychology, Universitas Airlangga, Surabaya	2019	Indonesia
Jiangsu Psychological Association (Development and Educational Psychology Special Committee)	2021	China
Jiaxing University	2012, 2014, 2016	China
Kyushu Sangyo University	2015	Japan
Nanjing Normal University	2021	China
National Chung Cheng University	2007	China Taiwan
National Sun Yat-sen University	2013	China Kaohsiung
Ochanomizu University	2008	Japan
Statistics.com The Institute for Statistics Education	2015	
Shanghai Foreign Language Education Press	2018	China
Shanghai Jiaotong University	2018	China
Springer Nature	2012, 2014, 2015 2016, 2023	German
SAGE	2018	
Social Media	2018	China
The US Meta Metrics Data Company	All the time	USA
The World Sports University, Macau, SAR, China	2023	China Macau
Singapore University of Social Sciences	2011, 2025	Singapore
Universiti Malaysia Sabah	2017	Malaysia
University College MAIWP International	2024	Malaysia
Vietnam National Institute of Educational Sciences	2022	Vietnam

Pacific-rim and beyond. Just as Professor Mok said, "Instead of being confined within the Pacific-Rim regions and countries, participation of scholars from Europe will make PROMS more' International". The second one is sort of exciting: More Rasch-based research by Chinese scholars has entered into the world. The papers and keynote speeches presented at the conference reflect both the new trends and status quo of the research and application of Rasch model in specific fields. The third highlight is theoretical: PROMSers have come to be aware that there does exist a little difference between Rasch model and the one-parameter of IRT. The fourth one is fundamental. The working units home and abroad have attached great importance to and have given support in one way or another to PROMS. This is listed in alphabetical order in Table 6.4.

6.5 Summary

This chapter introduces PROMS in four parts. The first part gives the introduction to PROMS in general from the first PROMS Conference in Malaysia, the first PROMS in China Mainland, the online PROMS conferences during the COVID-19 pandemic period to the first PROMS on-site conference after the COVID-19 pandemic period. Apart from this, PROMS Tenet is presented. The next part focuses on the features inherent in PROMS preconference workshops over the past 20 years. The third part introduces the themes and keynote speeches at PROMS conferences over the past two decades. The final part concentrates on both the development and the impact from PROMS in the growth of the Rasch-based publication, conference proceedings, Rasch monographs and future development. The whole chapter has provided a complete picture of PROMS currently being practiced. Therefore, to conclude, the author firmly believes that today anyone seriously interested in research field of psychometrics, social sciences or measurement will find PROMS such as an international symposium and related workshops to be an excellent source of information about the application of the Rasch Model.

Notes

1. Prof. Magdalena Mo Ching Mok is professor of evaluation and assessment at the Department of Psychology and the director of the Evaluation Research Center, The Hong Kong Institute of Education (HKiEd) (Now The Education University of Hong Kong). She also serves as Professor of Evaluation at National Taichung University of Education, and Editor in Chief of the British Journal of Educational Psychology. Formerly served as senior lecturer at City Polytechnic University in Hong Kong. Prof. Mok's research scope includes self-directed learning, evaluation

reform, Rasch model data analysis, and assessment to promote learning. Prof. Mok studied under Ben Wright and is a senior Rasch research expert and scholar.
2. EdSphere is an educational technology platform (also a brand name) designed and developed by Next Education India Pvt. Ltd. to enhance learning experiences by providing personalized and adaptive learning solutions. EdSphere incorporates data-driven insights to tailor educational content to individual student needs, helping educators track progress and identify areas where students may require additional support. EdSphere aims to improve educational outcomes by making learning more engaging and efficient for both students and teachers. However, specific features and offerings can vary, so it's important to refer to the platform's official resources for detailed information.
3. Ray Adams, a prominent figure in the field of educational measurement and psychometrics. Known for his work on large-scale assessments and his contributions to the development of measurement models. Ray Adams has been associated with the Australian Council for Educational Research (ACER), where he has played a significant role in research and development projects related to educational assessment. As of the author's last update, Ray Adams was not specifically tied to a university but rather was working with ACER. Please note that affiliations can change, and it might be useful to check the latest information from reliable sources for the most current details on his professional status.
4. To find more specific information about the workshops, interested readers might consider looking into the conference's official records or archived materials from that year.
5. These pre-conference workshop runners, among others, shared their insights and provided training on various aspects of objective measurement and Rasch modeling during the pre-conference workshops. However, specific details about the exact lineup of presenters for the PROMS2008 conference might require consulting the official conference documentation or proceedings for precise information.
6. For detailed information about the topics covered in the three workshops, interested readers may want to consult archived conference materials or official documentation from the event organizers.
7. Prof. Harris Shah Abd Hamid is the host of PROMS2024, Kuala Lumpur and the professor of University College of MAIWP International & Universitas Negeri Malang, Malaysia.
8. Dr. Yan JIN is a professor of Applied Linguistics at the School of Foreign Languages, Shanghai Jiao Tong University. She has been involved in the design and development of the College English Test since 1991 and served as Chair of the National College English Testing Committee of China since 2004. She has been involved in many research projects related to test design and development and has published many articles in academic journals. She is currently on the editorial board of international journals such as Language Testing, Classroom Discourse, and Chinese journals such as Foreign Languages in China, Foreign Language World, Foreign Language Education in China, Contemporary Foreign Languages Studies, etc. She is also editor-in-chief of Language Testing in Asia.
9. Eric Wu is a senior software developer in Department of Psychology at UCLA, specializing in computational statistics since mid-1980s. Wu is instrumental in developing the renown structural equations modeling program EQS with Dr. Peter Bentler. His work is EQSIRT, a comprehensive IRT program under the grant from National Cancer Institute of NIH.

10. Dr. Alan Urmston works as an academic in the English Language Centre at Hong Kong Polytechnic University. He is the team leader responsible for the Graduating Students Language Proficiency Assessment (GSLPA) and heads the inter-university collaboration which developed the DELTA.
11. Dr. Michelle Raquel has been a driving force behind the DELTA since it began as the Tertiary English Language Test. Michelle is a Language Instructor in the English Language Centre at Hong Kong Polytechnic University and provides the Rasch analysis which supports the Diagnostic English Language Tracking Assessment (DELTA) online.
12. Dr. Gwendoline Guan is a member of the DELTA team and is currently undertaking her doctoral studies under the supervision of Prof Trevor Bond. Gwen is conducting an assessment-based study of DELTA listening sub-skills using Rasch Measurement in order to enhance students' EFL listening development through diagnosis.
13. Prof. Xiaoting Huang studied quantitative methods and evaluation at University of California Berkeley (UC Berkeley) under the guidance of Prof. Mark Wilson and graduated with Ph.D. in Education in 2010. Huang got her MA degree in Education at UC Berkeley in 2007 as well. Huang's research interests cover measurement: test construction and validation (including item development. item analysis, reliability and validity analysis), item response modeling (unidimensional and multidimensional item response modeling), and statistics, especially in application of hierarchical/multilevel modeling in large-scale educational data analysis value-added models to evaluate school efficiency.
14. GaoKao is the abbreviation expressed in Chinese language, the shortened form whose literal meaning is "National Entrance Examination for Higher Education".
15. Dr. Yaru Meng, professor at School of Foreign Studies, Xi'an Jiaotong University, China, Council Member of China Association for Language Testing and Assessment. Her research interests cover Language Assessment, Cognitive Diagnostic Assessment, Dynamic Assessment and Language Teaching Technology.
16. Dr. Zouth Fong Chieng is of Faculty of Education, University of Malaya, Kuala Lumpur, Malaysia and working as an expert of language testing in Malaysian Examination Authority under the Ministry of Education, Malaysia.
17. Dr. Bambang Sumintono, the board member of PROMS, graduated from the program of Master of Education, Flinders University, Australia in 2001and with PhD. from Victoria University of Wellington, New Zealand in 2007. Previously Prof. Bambang worked as a lecturer at Universiti Teknologi Malaysia and Universiti Malaya, both in Malaysia. Currently he is a lecturer at the Faculty of Education, Universitas Islam Internasional Indonesia, Depok, Indonesia. His research interests are in educational administration and leadership, educational assessment, and Indonesian education development. He wrote two textbooks about application of Rasch modeling for social sciences research in Indonesian language. Bambang is also an editorial board member of: Educational Assessment Evaluation and Accountability, School Leadership and Management, and Leadership and Schools Policy. Dr. Bambang can be reached at: bambang.sumintono@uiii.ac.id
18. Dr. Haniza Yon, the founding President of the Malaysian Psychometrics Association (MPA), is currently serving her eleventh presidential term. She is a psychometrician, advocating fairness and high standards in testing. She has been working in psychometrics for more than 25 years, including as a psychometrician at ETS, USA, as a Vice President at Integrated Knowledge Systems, Illinois,

USA, as a Principal Researcher in the Artificial Intelligence Lab and a Director of the Psychometrics and Cognitive Analytics Labs at MIMOS Berhad, Malaysia. She received her PhD in Measurement and Quantitative Methods from Michigan State University, USA and served as an adjunct professor and fellow at the School of Quantitative Sciences, School of Economics, Finance and Banking UUM, the Educational Research Lab, UPSI, the Laboratory for Statistics and Computation, ISLA, Portugal and as a research fellow at the Psychometrics Centre, Cambridge University, UK. Haniza can be contacted at haniza.yon@globalpsytech.com.

19. Dr. Jing-Jyi Wu is the Chair of TERA & PROMS2013, and the member on local organizing committee of PROMS 2013 and the emeritus professor of National Chengchi University, Taiwan, China.
20. The "Eight-legged Essay" (*bagu essay*) was a highly stylized and rigid form of examination paper that became the standard in Chinese civil service exams from 1487 to 1901. These exams were the primary method of recruiting officials into the bureaucracy. Key Features of the Eight-legged Essay include:

    - Structure. The essay consisted of eight main headings and was limited to 700 Chinese characters.
    - Content. The topics addressed in the bagu essay were confined to the Nine Classics of Confucianism and had to be discussed in a certain set manner.
    - Purpose. The eight-legged essay was used as the chief yardstick in measuring a candidate's literary attainments.
    - Criticism. The format was often criticized for valuing style over substance and having no bearing on a candidate's ability to govern.

    The eight-legged essay was finally abolished in 1905 by the Qing dynasty.
21. The Needham Question, posed by the British scientist and historian Joseph Needham, asks why modern science developed in Europe during the Scientific Revolution of the 16th and 17th centuries rather than in China, despite China's earlier technological and scientific advancements. Needham spent much of his career studying Chinese science and technology, and his work highlighted the many innovations that originated in China long before similar developments occurred in Europe.

    The question is significant because it addresses the broader issue of why certain societies experience technological and scientific breakthroughs at particular times. It has prompted extensive scholarly debate and investigation into the cultural, economic, political, and social factors that may have contributed to the different paths of scientific development in China and Europe.
22. Prof. Jue Wang Ph.D. is currently a professor at Department of Psychology of The University of Science and Technology of China and a board member of PROMS since 2023. She received her Ph.D. in Quantitative Methodology Program under Educational Psychology at The University of Georgia, and previously worked in Research, Measurement Evaluation Program at The University of Miami. Her research focuses on examining rater effects in rater-mediated assessments using Rasch measurement models and unfolding models. She has published in peer-reviewed journals including Educational Psychology Review, Psychology of Aesthetics, Creativity and the Arts, Educational and Psychological Measurement, Journal of Educational Measurement, and Assessing Writing. She has also

co-authored a book (with Professor George Engelhard) entitled Rasch models for solving measurement problems: Invariant measurement in the social sciences published by Sage as part of Quantitative Applications in the Social Sciences (QASS) series.

23. Dr. Xun Yan, currently an Assistant Professor of Linguistics, Second Language Acquisition and Teacher Education (SLATE), and Educational Psychology at University of Illinois at Urbana-Champaign, USA where he is also the coordinator of the English Placement Test (EPT). His research interests include the development and quality control of post-admission language assessments, scale development and validation, rater performance and training, assessment literacy for language teachers, and test score use in educational settings. His work has been published in Language Testing, Assessing Writing, Journal of Second Language Writing, and Asian-Pacific Education Researcher.
24. Professor Guanzhong Luo holds a doctoral degree in Mathematical Psychology and has served as the Director of Assessment Technology and Research at the Hong Kong Examinations and Assessment Authority (HKEAA) for nearly 15 years. Currently Luo holds a professorship at South China Normal University and Jiangxi Normal University, China. Luo's research focuses on psychometric models and parameter estimation algorithms for achievement and attitude measurements, and his publications and computer programs for test data analysis are widely used globally. Luo can be reached via gz.luo@iclound.com
25. Rate Fold, a less known program developed with this algorithm is available from Luo who can be reached via gz.luo@iclound.com

Bibliography

Bond, T.G., & Yan, Z. (2016). Exporting to China: The Future of a Genuine Collaboration with the West. In Zhang, Q. (Ed.), Pacific Rim Objective Measurement Symposium (PROMS) 2016 Conference Proceedings Rasch and the Future (pp. 45–46). https://doi.org/10.1007/978-981-10-8138-5

Bond, T.G., Yan, Z., & Heene, M. (2021). Applying the Rasch Model: Fundamental Measurement in the Human Sciences (4th Ed). New York: Routledge Taylor & Francis Group.

Bulut, O. (2021). eirm: Explanatory Item Response Modeling for Dichotomous and Polytomous Item Responses [Computer software]. Available from http://CRAN.R-project.org/package=eirm

De Boeck, P., & Wilson, M. (Eds.) (2004). Explanatory Item Response Models: A Generalized Linear and Nonlinear Approach. New York: Springer.

Hoijtink, H. (1991). The measurement of latent trait by proximity items. Applied Psychological Measurement, 15, 153–159.

Likert, R. A. (1932). Technique for the measurement of attitude. Archives of Psychology, 140, 5–53.

Lord, F. M. (1980). Applications of Item Response Theory to Practical Testing Problems. Hillsdale: Erlbaum.

Luo, G. (2001). A Class of Probabilistic Unfolding Models for Polytomous Responses. Journal of Mathematical Psychology, 45, 224–248.

Luo, G., Andrich, D., & Styles, I. (1998). The JML Estimation of the Generalized Unfolding Model Incorporating the Latitude of Acceptance Parameter. Australian Journal of Psychology, 50, 187–198.

Mok, M.M.C., & Zhang, Q. (Eds.). (2014). Constructing Variables. Book of Abstracts Vol. II. Journal of Applied Measurement. ISBN: 978-1-934116-10-4. Maple Grove, MN, USA: JAM Press.

Mok, M.M.C., & Zhang, Q. (Eds.). (2015). Constructing Variables. Book of Abstracts Vol. I. Journal of Applied Measurement. ISBN: 978-1-934116-11-1. Maple Grove, MN, USA: JAM Press.

Mok, M.M.C., & Zhang, Q. (2018). Introduction to Rasch Measurement. ISBN: 978-1-934116-13-5. Maple Grove, MN, USA: JAM Press.

Roberts, J.S., Laughlin, J.E., & Wedell, D.H. (1999). Validity issues in the likert and thurstone approaches to attitude measurement. Educational and Psychological Measurement, 59, 211–233.

Roberts, J.S., & Laughlin, J.E.A. (1996). Unidimensional item response model for unfolding responses from graded disagree-agree response scale. Applied Psychological Measurement, 20, 231–255.

Rost, J., & Luo, G. (1997). An Application of a Rasch-based Model to a Questionnaire on Adolescent Centrism. Chapter 26. In J. Rost & R. Langeheine (Eds.), Applications of Latent Trait and Latent Casith Social Sciences. Waxmann Verlag GMBH: Münster and New.

Smith, R.M. (1990). Theory and practice of fit. Rasch Measurement Transaction, 3, 99.

Steiner, S. (2021). "Better Measurement and Fewer Parameters! The True Value of Rasch over IRT", The keynote speech at PROMS2021, Nanjing, China.

Thurstone, L.L. (1928). Attitude can be measured. The American Journal of Sociology, 26, 249–269.

Van Shuur, W.H. (1984). Structure of Political Beliefs. Amsterdam: CT Press.

Zhang, Q. (2013). Towards International Practice of Language Testing in China. Keynote speech given at the PROMS2012. Jiaxing, China, August 6–9, 2012. In Zhang & Yang (Eds.), Pacific Rim Objective Measurement Symposium (PROMS) 2012 Conference Proceedings. https://doi.org/10.1007/978-3-642-37592-7

Zhang, Q. (Ed.). (2016). Pacific-Rim Objective Measurement Symposium (PROMS) 2015 Conference Proceeding. https://doi.org/10.1007/978-981-10-1687-5

Zhang, Q. (Ed.). (2018). Pacific-Rim Objective Measurement Symposium (PROMS) 2016 Conference Proceeding. Springer doi: 10.1007/978-981-10-8138-5

Zhang, Q. (2019). Rasch Model: Research and Practice in China. In Myint Swe Khine (Ed.), International Trends in Educational Assessment. Brill | Sense. Retrieved from http://catalog.loc.gov.

Zhang, Q. (Ed.). (2024). Proceedings of Pacific-Rim Objective Measurement Symposium (PROMS 2023) AP/Springer Nature. https://atlantis-press.com/proceedings/proms-23

Zhang, Q., & Yang, H. (Eds.) (2013). Pacific-Rim Objective Measurement Symposium (PROMS) 2012 Conference Proceeding. https://doi.org/10.1007/978-3-642-37592-7

Zhang, Q., & Yang, H. (Eds.) (2015). Pacific-Rim Objective Measurement Symposium (PROMS) 2014 Conference Proceeding. https://doi.org/10.1007/978-3-662-47490-7

Zhang, Q., & Zhang, T.T. (2014 [2015]). Rasch Model: Status Quo and Prospect in China. In Zhang & Yang (Eds.), Pacific Rim Objective Measurement Symposium (PROMS) 2014 Conference Proceedings Rasch and the Future (pp. 17–24). https://doi.org/10.1007/978-3-662-47490-7

Index

Note: Page numbers in **bold** and *italics* refer to tables and figures, respectively.

D

E

J

K

L

M

Q

R

S

T

For Product Safety Concerns and Information please contact our EU representative GPSR@taylorandfrancis.com
Taylor & Francis Verlag GmbH, Kaufingerstraße 24, 80331 München, Germany

www.ingramcontent.com/pod-product-compliance
Ingram Content Group UK Ltd.
Pitfield, Milton Keynes, MK11 3LW, UK
UKHW022133190926
475702UK00013B/1789
* 9 7 8 1 0 4 1 1 2 5 9 5 2 *